I0823058

Grizzly bear
(Ursus arctos horribilis)

Also by This Author

Mountain

The Frozen Worlds

Animals Lost and Found

How to Talk to a Tiger ... and Other Animals

To Jess, Mark, June, and Huck

"For the strength of the Pack is the Wolf,
and the strength of the Wolf is the Pack."
—*Rudyard Kipling,* The Jungle Book

North American mountain lion
(Puma concolor couguar)

GRIZZLED

Love Letters to 50 of North America's Least Understood Animals

JASON BITTEL

FOREWORD BY JOEL SARTORE

WASHINGTON, D.C.

CONTENTS

Mexican spotted owl
(Strix occidentalis lucida)

Grizzly bear
(Ursus arctos horribilis)

FOREWORD

I'VE BEEN TAKING PORTRAITS of animals for the National Geographic Photo Ark for almost 20 years now. I've taken thousands and thousands of images, every one intended to invite the world to look more closely at the animals with whom we share this planet. Over time, these photographs have been used around the world in myriad ways, in magazines and books and, of course, online. They've also shown up on billboards, been used as tattoos, and even been projected onto the Vatican.

But the way they're used in the book was completely new to me.

Here we see the photos turned into paintings, so to speak, thanks to technology and artist Lisa Monias's steady hand. How interesting to come upon a whole new way of seeing the animal faces I've long since grown accustomed to, and a new opportunity to get the world to finally pay attention to them.

At its heart, the Photo Ark is a decades-long public education campaign to get people to care about the current state (and upcoming fate) of all creatures great and small. And the same could be said about Jason Bittel's book, which you're about to dive into. I'd love to know how many adults and children will finally step inside the tent of conservation thanks to these pages.

A sense of wonder and hope—that's something Jason and I have very much in common. Both of us know that most of nature is still out there, patiently waiting for us to do the right thing: Save vast amounts of habitat, reduce carbon footprints, and be careful about our use of pesticides, herbicides, plastic, and so much more.

But first, we must get people to fall in love, and this is something that we both do by telling stories—a strategy as old as humankind.

And so, in we go, with page after page revealing life on Earth. Three entries in this book have a special place in my heart.

First, there's the bald eagle, a great success story. The pesticide DDT had caused raptor numbers to plummet decades ago because it thinned their eggshells. Once we got rid of this chemical, the birds began to raise more young, and today the eagle, as well as the peregrine falcon, has recovered significantly.

Second, there's the monarch butterfly, a poster child for invertebrate conservation if ever there was one. Large, colorful, and one of the first insects we recognize as children, the monarch is in trouble because there are not enough nectar-bearing native flowers, especially milkweed, to sustain its life cycle along its migratory route. Converting our backyards and office lawns into native plant gardens would help tremendously—and it's a joy, saving nature literally in your own backyard.

And finally, though I try not to have favorites, the animal that comes closest is the star-nosed mole. It lives in the soil near bodies of water and is nearly blind, but that doesn't stop it. Come nightfall, it dives right in to hunt aquatic invertebrates and small fish by using its supersensory nose to find prey. So how does it not get lost? By blowing out and sucking back in hundreds of air bubbles a minute, smelling them, and memorizing its route all along the way. Absolutely amazing.

Every species of animal you see in this book has a great story to tell. But they all have something else in common: All are waiting for us humans to save the day, to finally care deeply about them and be inspired to save them and their habitats. It is not too late, but from here on out, saving nature must be intentional. It will take thought and effort and kindness if we are to preserve the world's oceans, prairies, and forests. The stakes couldn't be higher, for without them, there is no us.

I truly believe that saving nature can start with an imaginative book like this. After all, it worked for me, when my mother gave me a book about birds when I was in elementary school.

So here's to all the stories we tell—Jason, myself, and so many others—as well as to the species we learn them from. Long may they thrive.

—**Joel Sartore**
National Geographic photographer and founder of the Photo Ark

INTRODUCTION

DID YOU KNOW OPOSSUMS are pregnant for just 13 days, giving birth to blind, bald jelly beans? Or that deer antlers are actually temporary, disposable face organs that can grow at a rate of nearly an inch a day? Or that some fireflies mimic the flash patterns of other firefly species so they can catch and devour them?

These are just a few of the eye-popping animal facts scientists have taught me about North American wildlife over my career as a science writer—and there are hundreds more waiting for you within this book.

Just over a decade ago, I quit my job in advertising and started focusing all my writing on stories about the weird and wonderful creatures living all around us. I've eaten termites, caressed skunks, sniffed sloth turds, and caught bird-eating bats amid unexcavated Maya ruins.

Now this may not be everyone's idea of a dream job, but for me, it's been blackberry pie. Because each and every day, I get to learn new, mind-blowing details about wildlife, and then share those pearls of wild wisdom with the world.

Sometimes, that has meant huddling in the dark as 12-foot-tall bush elephants silently surrounded me. Other times, it's meant getting bluff-charged by a black bear or helping wildlife managers drag a half-rotted, half-eaten elk carcass out of the woods so they could investigate its cause of death.

Some days, it feels like all I do is write emails.

But whatever the task or story, there's always been one driving force throughout my science writing career: *wonder*. No matter how much I get to learn, no matter how many experts I interview or scientific studies I read, I will forever be in awe of how much there is to know about any given creature.

Right now, even as you read this, there are scientists dedicating their lives to the study of millipede mobility. To the life cycles of snails. To the mysteries of the Allegheny woodrat. I know, because I have talked with these experts, and I couldn't write the stories I do without them.

Of course, most of us can't spend all day every day investigating how woodpeckers work. We have other jobs to do, kids to take care of, soccer teams to coach, and any of a million other obligations that require our attention. And yet, just because you're now a teacher, carpenter, or emergency room doctor does not mean you have to give up your curiosity about the natural world.

It's for all of you that I've written this book—the ones who never stopped loving animals and wanting to know more about them.

Inside, you'll find easy-to-read chapters and jaw-dropping artwork about 50 animals that range throughout North America. Some of them you know, some of them you think you know, and maybe some of them you've never even heard of. But my hope is that after you've read this book, you'll be awestruck by every single one.

I WANT THIS BOOK to be for everyone—from the most experienced outdoorsperson, hunter, hiker, and fisher to the amateur birder, or even the person who'd rather see animals in a television documentary than on a trail. No matter your background or experience level, I aim to take you on a wonder-driven ride through North America's animal kingdom. From axolotls and armadillos to wolverines and yellowjackets, you will learn not just how big a critter is or what it eats—the facts a standard field guide might divulge—but also intimate, awe-inspiring details about how that animal fits into the world around it. You'll find out, for instance, how buffalo are pollinators, hummingbirds are warriors, and salmon help forests grow.

Each creature, from the most defiant wolverine down to the tiniest carpenter ant, is a grizzled survivor of ancient fates. Each is the product of countless waves of predatory interactions, environmental upheavals, and behavioral trial-and-errors. The gentle deer in your backyard would not exist as it does now without long-extinct dire wolves and saber-toothed cats that chased its ancestors. The yellowjackets hounding your picnic don't do so out of malice, but because they are black-and-yellow Valkyries on the hunt for flesh with which to feed their helpless young.

And for many creatures, humans have become a part of the story, too, in ways that are both good and bad. For instance, World War II and the shortages of insecticide it created led to the rise of eggshell-thinning DDT and, ultimately, the near extinction of bald eagles. But human compassion and ingenuity also recognized the danger in time and altered the bird's doomed trajectory before it was too late.

We'll talk about the chemical weapons of skunks, the sometimes surprising politeness of rattlesnakes, and why mountain lions are the largest cats on Earth with the ability to purr. You'll get an understanding of why manatees are not stupid, despite having brains without any wrinkles, as well as an explanation for why great white sharks don't eat *more* people than they occasionally do.

What you will not find in this book is an exhaustive list of everything we know about a given species. For one, we didn't have the room. And two, most people wouldn't want to read it even if we did.

You'll also notice that the groupings get a little weird. Some chapters cover individual species or subspecies, while others focus on entire groups. Likewise, reptiles are sectioned with amphibians, despite being separated by hundreds of millions of years of evolution. But those groups are often studied together under the banner of herpetology, so we've put them together here. Similarly, the invertebrate section includes many disparate lineages of critter, from arachnids and insects to cephalopods and cnidarians (jellyfish). I know, this is less than ideal, taxonomically speaking—but considering invertebrates make up more than 95 percent of all animals species on this planet, I'm just thrilled we could include as many spineless wonders as we did in this book!

On a similar note, given that there are many more than 50 different kinds of animals inhabiting the North American continent, I had to choose. And believe me, it was agonizing trying to decide which species to include. If I missed your favorite creature this time around, I apologize in advance. You could argue about where North America begins and ends, depending on whether you're looking at geographic features or political borders. For this book, we're considering animals that live in Canada, the United States, and Mexico, but not those in, say, Greenland or Costa Rica.

IN MANY WAYS, the idea for this book came from a conversation I once had with an evolutionary ecologist named Lee Dyer. Walking through the Costa Rican rainforest, trying to avoid stepping on bullet ants, Dyer told me something that lodged in my heart like a nematode.

Dyer studies the interactions between caterpillars and parasitic wasps. The wasps lay their eggs inside the caterpillars, which then hatch and eat the caterpillars from the inside out. What's more, some wasp species actually parasitize the wasps that parasitize the caterpillars. This is what scientists call hyperparasitoidism. And there may even be hyperparasitoids that parasitize the hyperparasitoids. All of this drama unfolding within the tiny, squiggly body of a caterpillar—so much wonder per square inch, it boggles the mind.

Anyway, Dyer told me that each interaction between host and parasite is like an epic poem. Most interactions, he said, remain undiscovered. And due to many factors—such as pesticides, climate change, and habitat destruction—these interactions are disappearing faster than we can describe them.

"They're poems that have never been read," said Dyer. "It's like burning down a library full of books."

To twist Dyer's metaphor just a bit, what follows is my attempt to reveal the epic poem hidden inside every animal. The evolutionary path of how it got here, or the biological innovations that let it do what it does. I'm talking about an animal's very essence. Its je ne sais quoi.

And I do mean *every* animal. This book is a love letter for the oft-maligned tarantulas, condors, and jellyfish just as much as it is for the ever popular orcas and hummingbirds. It's a pump-up speech for the opossums we sometimes hit with our cars and a power ballad for the ants that form assembly lines across our kitchen counters. And unfortunately, for some of the animals featured—I'm looking at you, axolotls and freshwater mussels—these pages may one day soon read like a heartfelt eulogy as their subjects disappear from Earth altogether.

In short, I hope this book serves as a recalibration for the ways we view and engage with the living things all around us. But at the end of the day, if for you, dear reader, it's just a book full of weird facts about mountain goats and Gila monsters, I'm okay with that, too.

Who knows, maybe learning more about these wild animals will even reveal something about the grizzled thing each of us carries around inside ourselves.

Either way, I hope you fall in love with that part of yourself, as I have with all the wild things within.

Southwestern bobcat
(Lynx rufus baileyi)

PART I

MARVELOUS MAMMALS

THE BEARDED BRINGER OF LIFE

AMERICAN BUFFALO

SPECIES: *Bison bison* **STATUS:** Near Threatened **RANGE:** Pockets of the American and Canadian West **SIZE:** Up to 6 feet 6 inches tall at the shoulder; weighs up to 2,000 pounds **LIFESPAN:** ~20 years

---

WITH A SHAGGY COAT OF BROWN FUR, a back hump like a grizzly, foot-long horns, and steam streaming out of its nostrils, the American buffalo looks like a creature that thundered out of the mists of legend.

Buffalo, also known as bison, can stand more than six feet tall and weigh a solid ton, making them the largest animals on the entirety of the North American continent. No polar bear or alligator even comes close. People go to Yellowstone National Park and worry about getting attacked by bears, wolves, and mountain lions, but year after year, buffalo injure more tourists than any predator.

If that surprises you, consider that these animals are notoriously unpredictable and can charge at 35 miles an hour. Buffalo spin around quickly, jump high fences, and swim across rivers that once swallowed many a covered wagon. What's more, American buffalo have evolved to confront their enemies as a group, creating defensive circles around their young with their formidable horns forming an outward-facing pike wall, like a medieval army in phalanx formation.

In North America, we use the words "buffalo" and "bison" interchangeably. But scientifically speaking, buffalo live in Africa and Asia and include species such as the Cape buffalo and water buffalo, while bison are another group of animals that inhabit North America, Europe, and the Caucasus.

This may seem like a confusing set of biological distinctions, but for the Indigenous North American peoples that relied on these animals, "buffalo" is the older and more accepted word, on account of early relationships with French fur trappers. The term comes from the French *bœuf,* meaning "beef."

"Of course, there are over 330 Indigenous languages on this continent, and we would have had our own word for the animal in each of those languages," explains Jason Baldes, Senior Tribal Buffalo Program Manager for the National Wildlife Federation and a member of the Eastern Shoshone Tribe. However, "buffalo" became such a universal term that many tribes have used the word in their surnames for generations. "We're not going to change our family names to 'Bison' just because scientists tell us it's the correct term," says Baldes.

Out of respect for the people who have lived alongside these animals longer than anyone else on the continent, I've chosen to use "buffalo" in this book. In a similar vein, no story about these animals would be complete without acknowledging that, not so long ago, European settlers nearly drove the American buffalo to extinction in an attempt to destroy the Indigenous people who wove these animals into every fiber of their being.

You've probably heard some version of the idea that Native Americans "used every part of the buffalo." But I never truly understood what that meant until Baldes told me that, to his people, buffalo were akin to walking, wallowing Walmarts.

Indigenous peoples used tanned buffalo hide to make everything from clothing and blankets to cradles, sweat lodges, and tipis. Buffalo fur filled pillows and moccasins and was braided to make rope. Tendons became bowstrings, hooves were rendered into glue, tails were turned to knife sheaths, and bones were used as shovels, splints, war clubs, and structural support for sleds. The buffalo's gall bladder provided yellow paint, while the urinary bladder served as a container for food and water. Buffalo dung could be burned in lieu of firewood or pounded into a liquid-absorbing powder for diapers.

Now, consider that there used to be something like 30 to 60 million American buffalo lumbering across this land mass. And not just on the Great Plains in the continent's center but across 22 major biomes, including forests, tundra, taiga, shrub steppe, prairie, and deserts.

As recently as the year 1500, you could have found buffalo rumbling across the shadow of Alaska's Denali at the same time as buffalo herds roamed south through the Mexican state of Durango. From the southern shores of the Great Lakes and the high plateaus beneath the Rockies to the edges of the Gulf Coast and the salt licks of the Appalachian Mountains, buffalo once called all these places home.

But then, in the blink of an evolutionary eye, humans systematically exterminated the great herds until, in 1884, just 325 buffalo remained.

One oft-touted reason for the slaughter was the lucrative market for buffalo parts, such as pelts and tongues. But there was also a larger, more insidious endgame taking place: The U.S. government embraced annihilating buffalo as a strategy to subdue the Native American peoples that depended on the animals. And unfortunately, it worked.

Today, buffalo inhabit less than one percent of their former territory. The vast majority live behind fences like livestock. What's more, very few of those animals are actually genetically pure buffalo, because in the late 1800s and early 1900s, most of the remaining populations were crossbred with cattle. This means that while it seems like we've saved buffalo from extinction, because you can buy their meat in fancy restaurants and at the grocery store, what we have left is little more than a ghost, a shadow, a token of its former self.

"They don't exist in large numbers on large landscapes," says Baldes. "Buffalo are ecologically extinct."

Of course, Baldes and his colleagues are trying to change that. Not just for his people, but for the North American continent itself.

The Buffalo's Lost World

With each thunderous step, each shake of their beards, each wallow in the dirt, buffalo remake the world around them.

Buffalo are what's known as a keystone species. Like the center block that holds up an arch in masonry, keystone species have an outsized role within their communities. They support other creatures by providing food, shelter, or opportunities.

Buffalo have spade-shaped hooves that cut into the soil as they plod. And while the footsteps of a single animal might not do much cultivating, the stamping of a herd hundreds of thousands strong once created miles-wide tracts of tilled soil that stretched horizon to horizon. In all that disturbed dirt, seeds took root—seeds that fell out of a bison's gnarly tangle of fur as it shook off flies or laid down for a wallow.

Also known as dust-bathing, wallowing helps the animals shed hair and covers them in a layer of soil that protects against biting insects and ticks. This buffalo behavior also plays a social role, and sometimes buffalo seem to do it just for fun. Whatever the reason, wallowing can create huge, hardened pits in the earth the size of an inverted pitcher's mound.

While the buffalo like their wallows dry, the more they use these areas, the wetter they become, because as the soil compacts, the more water it can hold. Buffalo abandon muddy wallows eventually, but wetland plants, butterflies, and even salamanders move in, boosting local biodiversity.

Historically, buffalo would have maintained upwards of 100 million active wallows—a renewable natural resource that mostly evaporated alongside the great herds. But amazingly, hints of the wallow ecosystem linger on.

Baldes has discovered that relic wallows untouched by a buffalo for at least 130 years still show signs of greater plant diversity than surrounding grasslands. Yarrow *(Achillea millefollium),* arnica *(Arnica angustifolia),* and bluebells *(Mertensia paniculata)* prefer to grow in wallows, and each are considered cultural plants with medicinal properties, notes Baldes.

Another underappreciated role assumed by buffalo is something more commonly associated with bees and butterflies. "Buffalo are actually pollinators," says Baldes, because their big, beautiful beards get dusted with pollen as they wade through tall grasses. Birds also snatch tufts of buffalo fur and weave it into their nests as insulation.

Like African bush elephants, buffalo keep forests in check. By plowing up soil with their hooves and thrashing their horns against small trees and shrubs, they actually prevent trees from overtaking open habitats, such as plains, prairies, and savannas—each just as important to the planet as forests. And doing so creates homes for all kinds of plants and animals that have adapted to open spaces, including endangered species, such as the Karner blue butterfly.

Every time a buffalo goes to the bathroom, it nourishes the plant community. A single buffalo can unleash up to three gallons of urine and dung each day, and every squirt and chip deposits nitrogen, phosphorus, calcium, sulfur, and magnesium into the soil.

Unbelievably, just one buffalo chip can support up to a thousand individual insects of more than 300 species. And those buzzing insects provide food for turtles, bats, birds, and innumerable other creatures. Dung beetles roll up pieces of buffalo dung and move them around, actually dragging them belowground where they can support baby dung beetles. Whatever doesn't get eaten fertilizes the soil.

If you somehow aren't yet standing up and singing "Circle of Life," know that scientists now believe removing this keystone species led to such catastrophic damage of grassland ecosystems that it eventually contributed to the Dust Bowl of the 1930s—a human-caused ecological disaster that claimed 7,000 lives and left two million people homeless.

Fortunately, scientists like Baldes continue fighting to change the buffalo's fate, both by returning wild, genetically pure buffalo to tribes that want them, but also by studying the gargantuan impact these creatures once had on the natural landscape.

Who knows what more we'll discover about the American buffalo and its role as life-bringer as the modern herds multiply, age, and expand?

THE NIGHT HOWLER

BOBCAT

SPECIES: *Lynx rufus* **STATUS:** Least Concern **RANGE:** From southern Canada to Mexico
SIZE: Up to 2 feet tall at the shoulder; weighs up to 33 pounds **LIFESPAN:** Up to 15 years

---

IT WAS JANUARY IN HILLSBOROUGH, North Carolina, and a young girl was sitting in a pigpen waiting for her sow to give birth. It's quiet, tedious work watching a restless animal paw about the straw, thinking the time is here, only to have your hopes dashed again and again as the moon rises and the rest of the world sleeps. But then an unrelated sound ripped through the night—a wild, high-pitched screech, like a woman somewhere beyond the tree line screaming in agony.

"It was like *The Shining,* farm-style," Imogene Cancellare remembers about those first few moments alone in a darkened barn. "I was like, 'Oh my god, this is a living nightmare.'"

In the years since, Cancellare has become a conservation biologist with the global wild cat conservation organization known as Panthera, and she knows now exactly what made that haunting, night-splitting wail. "That is a female bobcat indicating that she is in heat," she says, laughing.

In bullfrogs and blackbirds, it's the males who sing to attract a mate. But bobcats are loners, staying away from their own kind until females are in estrus (or heat). And while this important life event can take place several times a year, it usually only lasts for an average of five to 10 days at a time. So when the female bobcats sense their baby-making time coming on, they shout it from the rooftops so all the fellas nearby can hear. An actual catcall, but in reverse.

Female mountain lions produce a similar sound, says Cancellare, but to her ears, it's a little softer than the screech of a bobcat, despite the fact that mountain lions are much bigger. "I have no idea why. I just think everything about the bobcat is larger than life," she says. "So that I-sound-like-I'm-being-murdered, come-on-down-for-a-good-time call? It's very bobcat-specific."

Despite that alarming call, plenty of people confuse bobcats with house cats. And Cancellare says the misidentification goes two ways. When most of us see a house cat at the edge of a field or in a blurry trail cam photo, we assume it's a bobcat. But when we see a bobcat, we assume it's a mountain lion.

In the northern part of the continent, you can also add Canada lynx to the equation, and in the southern part there are ocelots and even jaguars and jaguarundi. Most of us will never see any of these creatures, because they have been honed by evolution for stealth. But they are out there.

"Everyone thinks that bobcats are supposed to be kind of little," says Cancellare. But male bobcats in some areas can weigh nearly 40 pounds, which puts them on par with a small border collie.

Another thing adding to the confusion is the fact that bobcat markings can differ depending on region, with animals in the Southwest displaying a sandy brown coat, which looks completely different from the bobcats of Florida and California, which are densely spotted. "They just have all these beautiful rosettes that are really similar to jaguars," says Cancellare of the bobcats in those localities.

Also, there's this thing known as Bergmann's rule, which states that animals tend to get bigger toward Earth's poles and smaller toward the Equator. This means bobcats in Saskatchewan and South Dakota are considerably larger than bobcats in Oaxaca. "And so you have all this variation. It's just incredibly confusing to people," she says.

In general, though, there are just a few things that can help you tell a bobcat from an alley cat, cougar, or other feline lookalike. Up top, look for tufted ears and a flared facial ruff, a bit like feline mutton chops. And down below, keep an eye out for a tail that looks comically short. A bob, if you will. The bobcat's bobtail can be up to eight floppy inches long, or as short as two stubby inches.

We might know bubkes about bobcats, but these medium-size predators are literally all around us. Except for a chunk of territory south of the Great Lakes and the northernmost reaches of Canada and Alaska (where the larger lynx rules), bobcats can be found pretty much everywhere in North America. In fact, bobcats are the most widely distributed native felid on the continent. This despite centuries' worth of antipredator attitudes, active pressure from hunters and trappers, and more human developments and ecosystem impacts every single day.

So, how has this chop-tailed wildcat managed such resounding and continued success? Well, you might boil it all down to one word: attitude.

Bobcattitude

"I love bobcats, because of how bad their attitudes are," says Cancellare. "And I know that's humanizing them a little bit, and we try not to anthropomorphize wildlife when we work with them. But having observed bobcats in traps, having tracked and collared bobcats, and having released so many bobcats, I think everything about them is just incredibly audacious."

With soft, padded feet, bobcats can move silently through the underbrush until they come to "hunting beds," or lookout points along well-traveled prey pathways. Then, these sit-and-wait assassins simply bide their time until a rabbit or vole wanders through, at which point they lunge forward and sink their retractable claws into its flesh. When prey are already in sight, bobcats can stalk with speed as unnoticeable as that of a glacier. In one observation from South Carolina, scientists timed a bobcat's crouched pursuit and found that the predator took 13 minutes to move just three feet closer to its potential dinner.

With this skill set, the bobcat is able to prey upon an impressive array of wildlife, across a nearly endless mosaic of habitats. "Bobcats thrive in swamps and mountains and arid deserts and in urban environments," says Cancellare. While Canada lynx hunt snowshoe hare almost exclusively, bobcats take a more cosmopolitan philosophy when it comes to their menu.

"A bobcat would eat a snowshoe hare, but they're also going to eat rats. They're going to eat lizards, insects, squirrels, and all manner of things," says Cancellare. Bird eggs, fish, mice, voles, beavers—you name it, and a bobcat will maim it.

Indeed, bobcats will sometimes even take down prey many times their own size. "There's a ton of anecdotal evidence—from regular people, to biologists, to photos on the internet—where all of a sudden there's a deer running down a hill and riding on the back of it, on its neck, is a bobcat," says Cancellare. We're not just talking fawns—though bobcats do pick off deer and even pronghorn fawns with surprising regularity. And it's not just injured deer or deer stuck in deep snow—though Cancellare imagines those are regularly occurring scenarios, too.

No, this cat, which stands no taller than your knee, takes down full-grown does and bucks with a bite to the throat. "I think it's completely bonkers that they do that," says Cancellare. "But it certainly happens."

Bobcats can climb trees to escape predators, jump 12 feet into the air to snatch birds, and even swim when the occasion arises. Studies of genetic relatedness reveal that not even the Mississippi River gets in the way, says Cancellare.

One time, a bobcat was even filmed dragging a three- to four-foot-long shark out of the ocean in Florida.

"I just think everything about them is so ecologically interesting. They're absolutely adorable, but they don't need anybody's help," says Cancellare. "They're really secretive, really efficient little powerhouses in their environment."

But even bobcattitude isn't enough to save these creatures from something people use all the time—rat poison. Many rodenticides contain anticoagulants that cause rodents to bleed internally and die. But they don't do so right away. And this means mice, rats, chipmunks, and other small creatures eat those poisons and then run out into the world, where they can get picked off by owls, hawks, coyotes, mountain lions, and the like. (By the way, rodenticides frequently poison pets and people, too.)

Unfortunately, the more predators are exposed to anticoagulants, the more these toxins build up in their bodies—sometimes to catastrophic effect. For instance, one Panthera-led study found that when rodenticides get into bobcats, they short-circuit the cats' immune systems and leave them susceptible to a lethal skin disease known as mange.

"So it's a lot better to have a bobcat in your yard taking care of the rats than it is to put rodenticide down and poison your bobcats, and your owls, and anything else that's already out there trying to get rid of those rodents for you," says Cancellare.

And though you may never set eyes upon the local bobcats you benefit, maybe one day you'll hear the night-splitting shriek of a bobcat in heat—bobcattitude incarnate—and know that it's all been worth it.

THE UNDEFEATED

COYOTE

SPECIES: *Canis latrans* **STATUS:** Least Concern **RANGE:** Most of North and Central America
SIZE: Up to 2 feet tall at the shoulder; weighs up to 50 pounds **LIFESPAN:** Up to 14 years

---

BEHOLD, THE MIGHTY COYOTE—an animal humans have been trying to get rid of for centuries, but which we have inadvertently managed to make bigger and more prevalent than ever.

Pronounced either "kai-OAT-ee" or "KAI-oat," coyotes are canids, in the same family as wolves, jackals, foxes, and even our pet pooches. The word "coyote" is itself rather ancient, hailing from the Aztec word for the animals, *coyotl*. And while there's a lot of variation across the coyote's range, in general, the animals are gray with subtle bands of black, cream, and hints of red. Size-wise, coyotes tend to be about half the stature of gray wolves, and their snouts and ears are usually longer and pointier.

Maybe you haven't yet glimpsed a coyote in person or experienced their screechy chorus of bark-yips in the night, but chances are you will soon. Because coyotes are one of the few species out there that are actually *expanding* their range in this human-dominated world.

Before 1900, coyotes were mostly restricted to the western two-thirds of the continent, with the Great Lakes and Mississippi River their loose boundaries in the East. But since then, these courageous critters have loped into lands anew and are now common in the East from Newfoundland to Florida. They've also been on the move out West. Historically, the species would have wandered no farther north than British Columbia, but they now inhabit the Yukon Territory and even Alaska. Likewise, coyote range has zoomed south past Guatemala all the way to Panama. All told, coyotes have expanded their historical range by 40 percent since the 1950s. All of which means there are now more coyotes on this continent than ever before!

So, how did we get here?

Well, without getting too far afield, you should know that during the 17th and 18th centuries, Europe underwent a philosophical shift known as the Age of Enlightenment. And one of the basic tenets of this new way of thinking was that humans and nature were inherently separate, and that it was up to humans to conquer all the wildness of the world. And so, when all those Europeans started colonizing North America, predators of basically every shape and size received a black mark. After all, predators eat the food we want to keep for ourselves—livestock and wild game—and were generally thought to pose a threat to human life, liberty, and pursuit of happiness.

In many cases, the merits of this argument are dubious and complicated, but suffice it to say that the philosophy led to the general de-predatorization of the soon-to-be United States, especially. In the eastern part of the country, antipredator campaigns effectively obliterated gray wolves, red wolves, and mountain lions, while out West, apex predators such as grizzlies and gray wolves now inhabit only a sliver of their former glory.

Whole books have been written on this topic—and Dan Flores's *Coyote America: A Natural and Supernatural History* is a fantastic one—but let's just say certain groups of people have gone to considerable lengths to annihilate certain groups of animals, and coyotes are very much included in that lot. The difference is that where those people mostly succeeded with brown bears and gray wolves, they utterly failed with coyotes. And they failed repeatedly, despite the use of common weapons like guns, traps, and poison-laced bait, as well as ridiculous-sounding gadgets such as the Humane Coyote Getter.

Patented in 1934 and consisting of a .38-caliber cartridge and a cyanide capsule, the Humane Coyote Getter lured animals in with a scented rag and then, after a tug, exploded that poison directly into the animal's mouth. And what was "humane" about this, you're wondering? Let's just say it resulted in deaths that were terrible but swift, which actually was an improvement on other popular, slow-acting poisons used on varmints, such as strychnine or sodium fluoroacetate.

The U.S. government actually still uses a modern-day version of the Coyote Getter, known as the M-44 or cyanide bomb. The mechanics are basically the same as its predecessor, and while these devices can certainly kill coyotes, they also sometimes accidentally euthanize bears, foxes, opossums, raccoons, skunks, and even pet dogs. (The devices have injured people as well. In 1966, a Texas land surveyor was killed by a Coyote Getter. In 2017, an M-44 device sprayed cyanide into the face of a teenager and his dog, killing the pet and leaving the kid with chronic headaches.)

To give you a sense of just how strong antipredator sentiment was back in 1905, Montana's legislature at the time actually directed the state's veterinarian to catch and infect 200 wolves and coyotes with a skin-burrowing mite that causes a debilitating condition known as sarcoptic mange. Symptoms include itchiness, hair loss, emaciation, and sometimes death. The hope was that these animals would then spread the miserable affliction to their kin back in the wild, which they did. Biological warfare, in other words.

Add it all up, and between 1915 and 1971, the U.S. government recorded 5.4 million coyote kills. Even now, the U.S. eradicates some 400,000 coyotes each year as retribution for livestock losses and during hunting seasons and coyote roundups—events where people get together and kill as many coyotes as they can.

Yet, despite it all, these smaller cousins of the wolf haven't even flinched.

"The evidence shows us that, no matter what we do, we're not going to be able to kill off all the coyotes," says Christine Wilkinson, a carnivore ecologist and social ecological conservation scientist at the University of California at Berkeley.

But how? How have coyotes survived this all-out onslaught?

Well, while you can beat an individual coyote by many means, the coyote species simply cannot be killed by conventional weapons.

A Predator Built for the Modern World

Wolverines and gopher tortoises are examples of solitary species, or loners. They spend most of their lives apart, only coming together to mate. On the other hand, wolves and crows are social species, meaning they're usually found in groups.

But coyotes? They can alter their life strategies at a moment's notice, morph their social structure to avoid extinction, and fill whatever niche they wander into, be it Mexican deserts, the Great Plains, or the streets of Queens.

"Coyotes are quite flexible in their sociality," says Wilkinson, who is also a National Geographic Explorer. "So they often live in these small family units, but they can sometimes live with large groups of unrelated animals that they collaborate with, and they can hunt solo."

Each situation allows coyotes to utilize different food sources. Whereas some animals are specialists—pandas mainly eat bamboo, osprey survive on fish—coyotes are generalists, capable of shifting to suit immediate situations and needs. On their own, they might prey upon small rodents, fruit, songbirds, and insects. But in a pack, coyotes can take down goats, sheep, deer, and even gigantic prey many times their own size, such as elk and moose—though for really big animals, deep snow and physically weak prey may be necessary for the pack to make a kill.

Of course, not being picky is itself a lifeline, and that elasticity allows coyotes to thrive in essentially every habitat, from deserts to tundra, plains to forests, suburbs to city centers. One time, a coyote made headlines after finding its way onto the roof of a bar just across the river from Manhattan.

Now, here's the wild part.

When coyotes stay together in packs, two things happen. First, typically only the dominant pair, or alpha pair, will breed. And second, the pack will defend their territory from other coyotes—both actively, with violence, and passively, with scent-marking and howling.

However, if that pack is forced to disperse—say, if someone goes and cyanide-bombs the alpha pair—then more of those lower-level coyotes will break off and become transients. They do not defend territories, but rather wander where the search for food takes them. (This ability to group up or go it alone is known as a fission-fusion society, and it's a trait that's shared by another clever mammal—*Homo sapiens*.)

Oh, and all those pack members that wouldn't have been breeding before? They start hooking up. What's more, their litter sizes increase significantly. Which means that sometimes, when you kill coyotes, you can actually wind up making more coyotes. "I think coyotes are just always kind of ready to rock 'n' roll," says Wilkinson.

Rather than looking at all these coyote life hacks as impediments to controlling and eradicating the species, perhaps it's time we start celebrating them for what they are—wild animals that have bested us after at least 200 years of war.

"Coyotes are playing this really integral role in regulating populations of other mesopredators, so things like skunks and raccoons," says Wilkinson. "They're also very likely benefiting our ability to regulate disease, especially in urban areas where they're rodent regulators."

In their native range, coyotes earned a place in the food web over millions of years of evolution. But if you're now hearing howls in the East and anywhere else these animals are new, just remember—people set the table by annihilating all the other big predators. So how can we blame these clever canids for pulling up a seat and tucking in?

THE TREETOP RAIDER

GEOFFROY'S SPIDER MONKEY

SPECIES: *Ateles geoffroyi* **STATUS:** Endangered **RANGE:** Central and southern Mexico; Central America **SIZE:** Up to 2 feet long not including tail; weighs up to 18 pounds **LIFESPAN:** Up to 47 years (in captivity)

DOWN IN THE JUNGLES common to the southern realms of North America, there's a wild primate that is criminally underknown. It weighs as much as a groundhog, swings through the canopy like a cartoon character, and, somehow does rather resemble the arachnid it's named after.

Meet Geoffroy's spider monkey.

With jet-black fur and extra-long arms, legs, and tails, these primates use all five of their lanky, grappling limbs to swing across the forest canopy at seriously impressive speeds—a mesmerizing movement style that even has its own name, brachiation. Geoffroy's spider monkeys are also epic communicators, using a range of both vocal and nonvocal signals, such as gestures and body postures, to warn their peers of intruders or invite each other to play.

When it comes to the primate world, lots of media focuses on gorillas, orangutans, and chimpanzees, all of which are obviously enchanting creatures. But despite much less attention from scientists, North American primates have all kinds of fun stuff going on, too!

For instance, did you know that monkeys have only been in the Western

Hemisphere for around 30 million years? That may sound like a long time, but it makes them relative newcomers compared with animals such as opossums or alligators, which have been evolving on this continent since the time of the dinosaurs. So, how did all these monkeys get here?

Believe it or not, all evidence points to the idea that the first monkeys to arrive here did so by first crossing the high seas. On a raft. From Africa.

Here's why such an improbable theory holds water. In this hemisphere, scientists have found precisely zero monkey bones or monkey teeth that can be dated to more than 35 million years of age. In other words, these animals have no North or South American evolutionary precursors. They kind of just appeared. What's more, the fossil evidence we do have points to three separate primate lineages, which means that whatever sort of dispersal event occurred to allow monkeys to colonize the Americas had to have happened at least three separate times (or different monkey species journeyed over together, Noah's ark–style, though this seems even more unlikely).

Since no fossils exist to suggest that monkeys trekked across the Bering Strait to North America or from Antarctica to South America, we're left to speculate that the little primates got accidentally swept up on a mat of vegetation, made their way across the Atlantic, and landed somewhere in Brazil, Panama, or the like. I know, it sounds bonkers. But back then, the Atlantic wasn't quite as wide as it is now, and this is the best theory we have.

Intriguingly, those intrepid monkeys *changed* once they got here. Take, for instance, their tails.

"Monkeys from the Americas have prehensile tails. Monkeys from Asia and Africa don't," says Gabriel Ramos-Fernandez, a biologist at the National Autonomous University of Mexico and a National Geographic Explorer. The word "prehensile" means "able to grasp," and that life-altering attribute has allowed spider monkey species to adopt a lifestyle almost entirely divorced from the ground.

"This species is basically born in the trees, and they die in the trees," says Ramos-Fernandez, who in collaboration with other colleagues has carried out the longest continuous study of spider monkeys anywhere. "The ground, to them, is just some kind of weird environment."

Some exceptions do apply. Young spider monkeys sometimes come to the ground to play, he says, and during the dry season, even older monkeys will descend to bathe or drink from a stream. But in general, the canopy is the safest place for these animals, which can brachiate away from nearly any attacker. On the ground, all those long, gangly limbs become a liability and leave the animals vulnerable to attack by predators such as pumas and jaguars.

And that's why it's so groundbreaking that Ramos-Fernandez and his co-authors have found one other condition under which spider monkeys will descend to the forest floor.

When Spider Monkeys March to War

Aggression is common in the animal kingdom. But it is also measured and calculated. Even mismatched affairs, like an eagle versus a murder of crows or a lone elk against a pack of wolves, carry risk of injury or death to the attackers.

Even more rare are examples of groups of animals squaring off against their own kind. In fact, one study scoured all the research we have and determined that most mammals are very unlikely to kill members of their own species. From bats to whales, intraspecific murder just isn't done on a regular basis. One surprising outlier, however, are the primates.

Primates kill other members of their own species at rates six times higher than the rest of the mammals. Yes, adorable lemurs, brilliant chimpanzees, wily white-faced capuchins, and even thoughtful-eyed gorillas have all been documented killing their own (not to mention humans—arguably the most violent animal on Earth). So, what makes our branch of the family tree so prone to conflict? Surprisingly, it might be linked to those big, beautiful brains.

According to the same study, species who were both highly social and territorial were most likely to cross the thin red line of fratricide.

In 2006, in collaboration with a researcher named Filippo Aureli and others, Ramos-Fernandez published the first evidence that spider monkeys, too, could be capable of organized, intergroup violence. In seven instances from the Otoch Ma'ax Yetel Kooh Reserve in the Yucatan Peninsula, the scientists watched as three to four males joined together to sneak into enemy territory and cause trouble.

"They would come to the ground and behave very differently, very silent, and make a single-file line and go into the neighboring group's range," says Ramos-Fernandez.

The monkeys took to *the ground*. A place spider monkeys do not usually go. Perhaps as a way to slip under their neighbors' canopy-focused radar.

In each raid, the spider monkeys moved cautiously, surveying the forest around them and apparently trying to zero in on sounds made by the neighboring group. Then, they moved *toward* those sounds.

Of course, wolves and other territorial animals will sometimes trespass in enemy terrain as they search for food, but they also avoid the rival group while doing so. Wolves will even kill other wolves during such raids, but again, such aggression is thought to be defensive. Remember, conflict is dangerous, and most mammals actively avoid it.

But the spider monkey behavior was very different: They seemed to be hunting for other spider monkeys, specifically. Even though some raids lasted as long as three hours, the males spent almost no time foraging or eating while raiding, which suggests the incursions weren't really about food. The scientists did not witness any killings, but several of the instances did result in violence—chasing, biting, and painful screams from the victims. One female was pursued so fast and hard that she jumped out of treetops and into a lake to escape.

With so few encounters to judge by, the scientists can't quite determine why spider monkeys perform raiding behavior. But the leading theory is that when the ratio of males to females gets too low in their own groups, they may go looking to court or even steal neighboring females. It's also possible that the raiders were trying to find and kill rival males that had become separated from their group.

While Ramos-Fernandez hopes to learn more about all of this by doing a kinship analysis for each group—that is, using genetics to see who is related to whom—one thing is clear: We only know about spider monkey raids because of the decades invested into studying these two groups. Ramos-Fernandez and his cohort have been following these same spider monkeys for 24 years—a feat that rivals some of the work done on chimpanzees, bonobos, and gorillas but has received much less attention.

There's likely so much more to learn about these clever, curious little primates! In fact, spider monkeys have already been found to practice traditions, or behaviors performed by one or a few spider monkey populations but not all of them. Some populations kiss each other on the cheek as a way of saying hello, for instance, while others rub ficus roots all over their chests, perhaps as a way to mark themselves and differentiate between groups. It's also curious that spider monkey traditions tend to skew toward social behaviors rather than food extraction, like termite-fishing and some other behaviors chimpanzees and orangutans have become famous for.

Like coyotes, spider monkey groups can split up and form back together, like beads of water on a window pane—a lifestyle scientists call fission-fusion, which is thought to be linked to many of the spider monkeys' complex behaviors. For one, such a dynamic requires tons of communication between monkeys, and that's not something every dragonfly, diamondback, or dingo can do.

If all of these attributes are starting to sound eerily like humans, you should know that scientists think so, too. From the fission-fusion dynamics, to the intergroup aggression, to the communication skills and variable traditions, Ramos-Fernandez says many researchers now believe spider monkeys are the most intelligent wild primate species in the Americas. What's more, they believe studying spider monkeys can help us better understand how our own ancestors rose up out of the savannas and built the world as it is now.

Not bad for a lineage that wouldn't even be here if not for the winds of fortune, right?

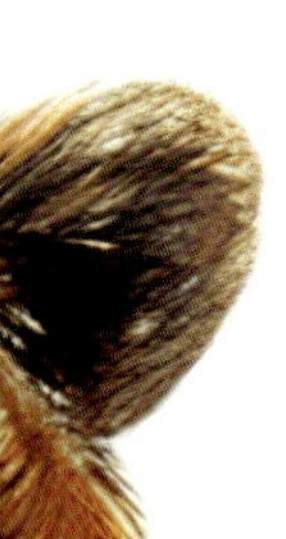

THE ORIGINAL BAD NEWS BEAR

GRIZZLY BEAR

SPECIES: *Ursus arctos horribilis* **STATUS:** Threatened **RANGE:** Northwestern North America **SIZE:** Up to 5 feet tall at the shoulder; weighs up to 800 pounds **LIFESPAN:** 25 years

---

THE GRIZZLY IS A SUBSPECIES OF BROWN BEAR—an offshoot so full of sass and vinegar that its scientific name, *Ursus arctos horribilis,* speaks for itself.

Black bears are smaller, faster, and better climbers. Polar bears tend to be bigger, heavier, and eat far more meat. But it's the grizzly's demeanor that makes it so infamously formidable. That and enough jaw strength to crack open a bowling ball like a pistachio.

As a bear biologist at the University of Alberta, Andrew Derocher works with grizzlies and polar bears for a living. And while polar bears have a reputation for being the world's most massive land predators, it's the smaller ursid that unnerves him.

"For two species that are supposed to be so genetically similar, they are just incredibly different," he says. (Polar bears actually split off from the brown bear species something like 343,000 years ago. As such, the two species can still interbreed and even produce fertile offspring, known as pizzlies.)

"When we approach them in the helicopter, the polar bears are laid back, don't seem to panic. But the grizzly bears just seem to go nuts with the helicopter around. Sometimes you chase them, sometimes they chase you."

"They jump at the helicopter," Derocher says. "They stand on their hind legs and try to pull you out of the air."

So are grizzlies just plain evil? Nah. They're simply oversize fuzzballs trying to make their way in the world—just like you and me. And scientists believe it's that dedication to self-preservation that is at the heart of the grizzly's bad attitude.

That old adage of not wanting to get between a mama bear and her cubs? It's not referring to black bears. When confronted by a predator or human, black bear mothers will beat feet or head for the nearest tree, because both they and their babies are equipped with short, curved claws that allow them to climb with the agility of kitty cats.

But grizzlies evolved in more open, treeless habitats where escape options were few. Grizzly claws are also long and straight, which is all good and well for digging and slashing, but not ideal for climbing. (Note: "Not ideal" does not mean grizzlies are incapable of climbing trees. They *can* climb. Just not as well as a black bear.) This means that when mama grizzly and her cubs come across a threat—which can include a big, bad male of her own species—flight might not be an option.

That means it's either fight, or die. Even if your enemy happens to be a helicopter.

What sets grizzlies apart is that this preset is so fundamentally at odds with most other wildlife. For example, think about the last time you saw an animal that did not run, swim, or fly away in fear the second you approached. Flight behavior is, by far, the norm.

But there's another thing that brings bears of both sexes into conflict with humans on the regular, and it's something you can probably relate to at least a little bit.

The Hunger

"Typically, when a bear wakes up from hibernation [in March or April], they actually continue to lose weight and don't start putting on mass for the next hibernation period until about July," says Kari Kingery, a wildlife biologist and member of the Confederated Salish and Kootenai Tribes in what is now northwestern Montana.

"And so they go into this stage called hyperphagia, which is that drive to consume as many calories as possible," says Kingery.

Sure, we all get hangry. But a grizzly fasts all winter while it hibernates. Some bears can lose up to 30 percent of their body weight before emerging in spring. If pregnant, female bears also give birth to up to four cubs during this period, which means they are then nursing on an empty stomach, for months, without so much as a snack.

So, what does a grizz fresh out of hibernation fill its maw with first? Given the bears' well-documented belligerence, you're probably assuming grizzlies survive on a diet of rattlesnakes, gunpowder, and backpackers, right?

"Here in the Mission Mountains, our bears have a strong diet of apples and fruit trees," says Kingery. "Oh, and also natural berries. Huckleberries being one. Chokeberries, serviceberries."

Apples and berries? You've got to be kidding me.

"Our bears also eat a higher proportion of bugs," says Kingery. "They're digging up ants and army cutworm moths."

Yes, friends—the gnarliest, most ferocious wild animal on this continent is fueled primarily by berries and bugs. By some counts, a single bear can gulp upwards of 40,000 moths each day.

Grizzlies also eat the following species in high quantities: grasses, sedges, forbs, roots, bulbs, wasps, and pretty little ladybugs. Sometimes, the bears even consume plain old dirt, likely as a way to acquire potassium, magnesium, and sulfur. Sure, bruins will gobble up everything from a moose on down to eggs in a bird's nest, if they can get it. But mostly, they're out there eating bugs, berries, salad, and soil. Not exactly the stuff of nightmares.

Interestingly, brown bear diets vary pretty dramatically across their range, as do brown bear sizes. In places such as Yellowstone National Park, grizzlies pack on pounds by feasting upon on whitebark pine nuts. The nuts are luxuriously fatty, and full of other nutrients like carbohydrates and protein. Best of all, red squirrels do all the work, gathering up to 15,000 whitebark pine cones and then storing them in neat little piles called middens. All the bears need to do is sniff them out and dig in.

Of course, brown bears that live along Canada and Alaska's coast are famous for feasting upon summer salmon, which migrate hundreds of miles inland to reproduce. An experienced bear might snag 75 pounds' worth of salmon each day, which translates to a glut of around 65,000 calories.

Now, a difference in diet also drives differences in bears. Salmon-eating brown bears routinely weigh more than 1,000 pounds, whereas Yellowstone National Park has never once recorded a grizzly that weighed more than 900 pounds. Likewise, Kingery says their largest grizzes may run just 500 pounds. And in Alaska's Kodiak Archipelago, another brown bear subspecies known as Kodiak bears *(Ursus arctos middendorffi)* sometimes weigh up to 1,500 pounds.

"The diet kind of determines what size the bears in that region are going to be," says Kingery.

Now, none of this is to say you should treat a grizzly that got fat on ladybugs any differently than one that got fat on caribou carcasses. When it comes to conflict, a grizzly is a grizzly is a grizzly, and if it feels threatened, a grizzly will cut through you like a piñata.

Grizzlies are the fastest bear species, with the ability to clock in at 35 miles an hour over short distances. For comparison, one of the fastest human beings to ever live—a phrase which will never be used to describe you or me—topped out at 27.5 miles an hour.

"So sometimes you have like three seconds," says Kingery. This is why many experts recommend defending yourself with bear spray rather than a gun.

Grizzlies have really thick skulls, says Kingery, and sometimes smaller caliber bullets—like those in a handgun—just ricochet right off them. What's more, as these animals jockey with each other for the best feeding spots and reproductive rights, grizzlies routinely endure flesh wounds that would put a human in the morgue. "Let's say you do injure the bear," says Kingery. "Its response at that point isn't going to be to run. It's going to be to continue that attack."

Actually, the best way to stay safe in bear country is to prevent an encounter in the first place. That means talking loudly so bears can hear you coming, especially where visibility is low, such as dense brush or thickets. Also, keep your dog on a leash. This prevents Sparky from running ahead where he might kick up a bruin and then come running back to you with the predator on his heels.

If that doesn't work though, bear spray has saved many a human life.

"The reason bear spray is so effective is because when they inhale that incredibly spicy pepper with that incredibly sensitive nose, it just kind of scrambles their brain," says Kingery. With an olfactory bulb in its brain five times bigger than our own, a grizzly's sense of smell exceeds that of a bloodhound. In fact, you could argue that smell is the species' most important sense.

The best part is, you don't have to be superaccurate with bear spray. Simply putting a pepper spray cloud between you and the bear means that huffing, puffing predator won't reach you without first coating its mucus membranes in red pepper oil and the tongue-tingling capsaicin it contains. And unlike the blast of a handgun, the searing pain associated with bear spray is basically the only thing proven to make a grizzly retreat (there is a catch, however—bear spray must be accessible at a moment's notice, not stored away in a pack).

To be abundantly clear, you do not want to tango with a grizzly. Ever. But if you do, consider that one weapon tries to hurt a creature that has been sculpted by evolution to withstand an unbelievable amount of physical damage, while the other takes the bear's highly specialized anatomy and turns it into a weakness.

So if you're going to put anything in a holster, please take Kingery's advice, and arm yourself with science.

THE MULTIMILLION-MOM ARMY

MEXICAN FREE-TAILED BAT

SPECIES: *Tadarida brasiliensis* **STATUS:** Least Concern **RANGE:** Western and southern United States and Mexico; parts of Central and South America **SIZE:** Up to 4.25 inches long with a 14-inch wingspan; weighs up to half an ounce **LIFESPAN:** Up to 18 years

---

IN THE TEXAS HILL COUNTRY just north of San Antonio, Bracken Cave opens out of the ground like a squinting eye. And every evening between May and September, visitors to this gigantic hole can watch as eight to 15 million Mexican free-tailed bats rise up out of the earth. Each bat is tiny, weighing no more than three board game dice apiece, but when they emerge at the same time, the mini-mammals turn the sky into a boiling tornado of furry flesh before flapping off into the night in an aerial river that meanders all the way to the horizon.

Bracken Cave is thought to be the largest single-species bat cave on the planet. So many bats in one place, it's beyond comprehension.

"Whether you're in the presence of fifty thousand bats or you're in the presence of two million bats, as a human being, you don't see the difference," says Rodrigo A. Medellín, an ecologist at the National Autonomous University of Mexico and National Geographic Explorer at Large. "You're overwhelmed by the sheer number of individuals that are flying around you."

Now, some might think this sounds like a nightmare. But Mexican free-tailed bats couldn't hurt a human if they wanted to. And they really don't want to.

While many people still associate these animals with vampires, out of more than 1,400 bat species known to science, just three eat blood—the common vampire bat *(Desmodus rotundus),* the hairy-legged vampire bat *(Diphylla ecaudata),* and the white-winged vampire bat *(Diaemus youngi).* This means more than 99 percent of all the bat species on Earth want absolutely nothing to do with us. (While occasional nibbles do occur, humans are not the preferred food source for *any* of the vampire bats.)

Bats also do not want to get tangled in your hair, even though TV and movies seem to think that's a thing bats do. It's not. I've visited Bracken Cave and stood speechless in the middle of the bat-nado—never once did any of those millions of mammals even so much as brush against me.

Finally, while bats can carry rabies, and in North America, they tend to account for more cases of transmission to humans than other animals, the numbers reveal that the actual risk is superlow. For comparison, rabies still kills about 59,000 people each year—mostly in Africa and Asia—but nearly all of those cases come from feral dogs.

On this continent, which is home to 374 million humans, annual rabies deaths can usually be counted on one hand. In a bad year, two hands. Statistically speaking, you're much less likely to be killed by a bat than you are an avalanche, deer, or tree.

Millions of Moth-Munching Mamas

It would be difficult to know this without peeking inside, but Bracken Cave is basically a giant subterranean nursery. And of the many millions of bats that visit it each year, almost all of them are female.

"The females all aggregate together in a maternity colony in the spring and summer and basically give birth around the same time," says Winifred Frick, chief scientist at Bat Conservation International, which owns and manages Bracken Cave.

Every year, Mexican free-tailed females pop out one pup each, which they stow away at the upper reaches of the cave in tightly packed clumps, called creches, when they leave to go out and hunt for insects every night. A single creche might have as many as 400 bat pups crammed into each square foot, and yet the mothers are able to find their own offspring each time they return using a combination of sound and smell.

The bats are so legion, their bodies literally raise the temperature of the cave. And that climate control? It's necessary for the babies to survive. With so many bats in one place, guano piles up and creates ammonia concentrations in the air that have been measured at around 250 parts per million—or a level considered fatal to humans if exposed for 30 minutes or more.

"When you're inside the cave, it is one of the most intense experiences," says Frick. "The air is supertoxic, you can't breathe, and yet they're in there, having their babies."

Did I mention that the floor of the cave is also a literal death trap? When scientists venture inside using respirators, any exposed skin is immediately attacked by flesh-eating dermestid beetles, which make a living on dead and dying bats. That includes pups that have fallen from their perch, which are skeletonized within minutes. This constant crawling, gnawing, and nutrient-shuffling keeps the guano ever-moving and pristine, like darkly colored dunes on the moon of a science-fiction film. Even the scientists' footsteps quickly disappear as the beetles scuttle to and fro.

While Bracken Cave is a critical site for Mexican free-tailed bat reproduction, the arrival of the bats at the cave each summer also sends a beneficial shock wave throughout the ecosystems of the Hill Country. This is because, as mammals, mother bats must generate loads of nutrient-rich milk for their babies. And to do that, they must feast.

Each night, a lactating bat has to eat around two-thirds of her own weight in insects to keep the baby's gravy train rolling. And lucky for us, some of these bats' absolute favorite midnight snacks include species that wreak havoc on our crops. Things like corn earworm moths, fall armyworms, beet armyworms, cabbage loopers, tobacco budworms, and cucumber beetles.

According to one of Medellín's studies, the mama bats save cotton farmers $688,000 each year in reduced pesticide sprays and damage to their bolls—and that's just for one crop found in the 30 square miles surrounding Bracken Cave!

Put another way, even if you estimate on the low side that there are 8 million mother bats in the sky on any given summer night, and each of those eats just 10 grams of insects, that would mean that the bats gobble up, in total, 80 tons of insects on a nightly basis. That's the equivalent of removing more than two blue whales' worth of bugs from the ecosystem. And this aerial cleansing occurs every single night!

Even when the bats are done raising their pups and they migrate back to Mexico and Central and South America, those ecosystem services go with them.

"So just picture for a second what would happen if we lose those bats," says Medellín, hinting at the hidden role these creatures play in pest removal on farms and other agricultural businesses virtually everywhere on Earth except Antarctica. "Project that out onto your morning coffee and your cotton clothes to the taco that I'm going to eat in a few minutes."

In the U.S. alone, the mammals save us around $3.7 billion dollars each year in pest control services. That's more than twice the gross domestic product of Belize.

By the way, that thought experiment about bats disappearing? It's not so hypothetical. As we clear forests to build houses, roads, and hardware stores, we carve up and destroy bat habitat. And all the pesticides we use to grow our crops, in turn, have killed a lot of the buggos bats eat. Furthermore, on top of those general threats, bats in North America have had to contend with a nasty little fungus that kills them while they hibernate. The resulting disease is called white-nose syndrome, and it has wiped out more than 6 million bats since 2006. Since Mexican free-tails don't hibernate where the fungus is present, they escaped the plague, but 12 other bat species weren't so lucky. For example, in northern long-eared, little brown, and tri-colored bat populations, nine out of every 10 bats died, says Frick.

Of course, bats are sometimes killed indiscriminately in the name of controlling rabies, COVID-19, and a number of other pathogens, despite the fact that culling bats only causes them to fan out, potentially spreading what you're trying to control.

All the more reason to embrace these magnificent mammals and the benefits they bring. Figuratively speaking, of course! If a bat wanders into your house, do not try to hug it. Just gently usher it back outside.

"Chances are, if you live in North America, you are going to be connected to the Mexican free-tailed bat," says Medellín.

Remember that on your next Taco Tuesday.

THE LITTLE ARMORED ONE

MEXICAN LONG-NOSED ARMADILLO

SPECIES: *Dasypus mexicanus* **STATUS:** Least Concern **RANGE:** From South America to the American Midwest and Southeast **SIZE:** Up to 2 feet 6 inches long; weighs ~12 pounds **LIFESPAN:** Up to 20 years

---

MAMMALS ARE A WEIRD BUNCH. They can come coated in fur, like a deer. They can have a sprinkling of spikes, like a porcupine. And they can even shoot chemical weapons out of their backsides, like a skunk. And then there's a whole lineage of tanklike mammals that waddle around with skin made out of bones—armadillos.

How is it that we humans belong to the same branch of the vertebrate family tree, and yet while armadillos are covered in a coat of armor, our skin can't even stand up to the sharp edge of a piece of paper?

The armadillo's secret lies in bony deposits known as osteoderms that are embedded in a layer of their skin ("osteoderm" sort of means "bone-skin" in Latin, and "armadillo" translates to "little armored one" in Spanish.) Only the size of a fingernail, each osteoderm fuses with those around it to create a series of overlapping shields.

While it may look like a shell, the armadillo's outer layer is "softer and more pliable than you'd expect, like a stiff leather," says Colleen McDonough, an armadillo researcher of more than 35 years. Oh, and this: "They feel warm." This is because the armadillo's protective coating is alive. Cut it, and it bleeds. Give it time, and the skin will even grow back over those osteoderms. In other words: bone-skin that can *heal.*

The other neat thing that sets the armadillo carapace apart from a turtle's or clam's: This armor is interlocking. One large, curved plate covers the animal's shoulders and neck (the anterior scapular shield) and another large plate protects the hips and butt (the pelvic shield). But in between are what's known as carapace bands, each of which overlaps and allows the armadillo to bend in the middle, not unlike a slightly hairy accordion. More osteoderms ensconce the rest of the body, shielding it from head to toe to tail.

"I've sometimes seen dogs trying to grab larger armadillos, and the teeth just slide right off the carapace. They can't really bite down," says McDonough, who is also the co-author of *The Nine-Banded Armadillo: A Natural History.* (New research has found the species previously known as the nine-banded armadillo to actually be four distinct species.)

Protected though they are, armadillos have very poor eyesight. And though they hear and smell well, you can actually sneak up on the critters in a way not possible with a raccoon or rabbit. However, if something—a stealthy researcher, perhaps—gets within range without tripping the armadillo's senses, and then that thing surprises the armadillo? Well, it can trigger a startle reflex that is nothing short of hilarious. Because when they are spooked, the little tanks launch themselves into the air like a surprised house cat—with stiffened legs outspread. Upon returning to planet Earth, the armadillo then rockets off into the brush. It may sound silly, but it works. When a predator sees its lunch catapult into the air like a Mexican jumping bean, the shock gives the 'dillo a head start into the nearest thicket.

You don't exactly have to be a pouncing puma to make an armadillo lift off, either. McDonough remembers one time when a large bird landed on a low-hanging branch near an armadillo she had been watching, and that swoop of wings was enough to send the armadillo nearly into orbit.

Unfortunately, there's one "predator" that makes armadillos startle more than any other—vehicles. And while springing into the sky might be an effective escape maneuver against a rattlesnake or gator, this leap actually causes the buggers to bash off of bumpers that might have passed harmlessly over them if they'd just hunkered down.

Strangely, while it's long been assumed that armadillos evolved their armor to protect against predators, no scientist has ever studied that relationship. It's true that the animals have few natural predators, though American alligators, coyotes, black bears, and even alligator snapping turtles eat them from time to time. But scientists don't just take things at face value. And so far, no one can actually say if armadillos get preyed upon more or less than similar-size mammals in the habitats they prefer, such as areas next to streams and rivers.

In fact, given that male armadillos do battle when courting for females, and that sometimes these duels result in nasty, deep scars on their living flak jackets—thanks to the Mexican long-nosed armadillo's seriously impressive set of claws—there's plenty of reason to believe that the carapace evolved to protect armadillos from each other.

Those fancy fingers also allow armadillos to burrow into the earth, where they spend most of their nonforaging hours. They also help the little armored ones tear into the nests of ants and other insects—some of the 'dillo's favorite foods.

While evidence for the evolutionary origin of armadillo armor is lacking, there is plenty to suggest that having skin made out of bone is rather handy for other reasons. For instance, armadillos can rush through thorns and thick underbrush without hesitation. The Mexican long-nosed species also appears to be blessedly devoid of fleas and ticks—at least in the quantities similar to what most other wild animals suffer. There simply aren't many places for a parasite to bite.

"I would say the carapace is probably multifunctional," says McDonough.

Quadruple the Fun

As conspicuous as the armadillos can be, scientists know surprisingly little about the animals, says McDonough. This is probably more because of the animal's nocturnal and solitary lifestyles, which makes studying them a bit of a chore.

As for availability, there are likely more Mexican long-nosed armadillos scuttling around than ever before. In fact, while the species was only seen as far north as Texas back in the mid-1800s, armadillos now inhabit everywhere from New Mexico, Oklahoma, and Arkansas to Tennessee, Florida, and South Carolina. There are even reports of armadillo sightings as far north as Minnesota, Wisconsin, and

South Dakota, though some are thought to have been transported illegally or accidentally by people. One estimate suggests the animals are moving north at a rate of about 2.5 miles each year.

Along with the other members of its genus, *Dasypus,* Mexican long-nosed armadillos are the only vertebrates known to consistently reproduce by way of quadruplets. And when I say "consistently," I don't mean "usually" or even "a majority of the time." I mean, every single time a Mexican long-nosed armadillo has babies, the infants will always be a group of four males or a group of four females—each individual an identical clone of the others.

Why do armadillos do this? Heck if we know! McDonough has read all the studies, and for her part, it seems most likely that there's some sort of hang-up in the female armadillo's anatomy where there is only enough room for one fertilized egg at the earliest stages of pregnancy. However, perhaps as a way to get around that anatomical quirk, evolution seems to have favored splitting that single armadillo embryo into four clones. More bang for the buck, so to speak.

Another armadillo oddity? Aside from us, the armored ones are some of the only other animals known to regularly contract Hansen's disease, or leprosy.

Caused by a bacterium called *Mycobacterium leprae,* leprosy leads to skin lesions and nerve damage in humans, especially in the extremities, presumably because the bacterium thrives in slightly colder environments (hands and feet are often cooler than your core). Armadillos have lower-than-expected body temperatures—as do their closest relatives in the superorder Xenarthra, sloths and anteaters—and it's thought this trait allows the bacterium to inhabit their entire bodies.

This is one reason why experts advise against handling wild armadillos or eating undercooked armadillo meat—which, by the way, is supposedly quite tasty and similar in quality to pork. If the armadillo is infected with the leprosy bacteria, there is a chance you can become infected, too.

Fortunately, only one-fifth of the Mexican long-nosed armadillos tested in the United States showed signs of leprosy bacteria infection (if that sounds like a lot, consider that Brazil's armadillos suffer from leprosy infection rates more than three times as high). Furthermore, of the 200 or so cases of leprosy documented in the U.S. each year, only 25 percent involve contact with armadillos. Not relieved?

Consider that 95 percent of the human population seems to be genetically resistant to the leprosy-causing bacteria, further reducing the odds of contagion.

Now, are you ready for the most peculiar part of the whole story? Even though armadillos sometimes give leprosy to humans today, we gave it to them first. That's because armadillos only exist in the Western Hemisphere, and a study of the leprosy bacterium's DNA shows that it did not exist in North America until Europeans brought it here. (If you're wondering how a human gives an armadillo leprosy—well, we don't actually know. It's possible the animals picked up the pathogen by digging in contaminated soils, though.)

The good news, depending on your perspective, is that armadillos rarely live long enough to feel the effects of the disease.

With cloned quadruplets and a coat of living armor we still don't fully understand, armadillos are not only one of the most exceptional animals on the North American continent, but they also remain one of the most mysterious. We should consider it an honor to dwell among them.

THE LOBO IN LIMBO

MEXICAN WOLF

SPECIES: *Canis lupus baileyi* **STATUS:** Endangered **RANGE:** American Southwest and northern Mexico **SIZE:** Up to 32 inches tall at the shoulder; weighs up to 80 pounds **LIFESPAN:** Up to 10 years

---

THE HOWL OF THE MEXICAN WOLF is one of the most endangered sounds on Earth. That's because there used to be thousands of these animals all across the American Southwest and Mexico, but today, only 257 wild Mexican wolves remain.

Technically a subspecies of the gray wolf, there are two easy ways to distinguish Mexican wolves from their canid cousins—size and color. Though gray wolves can weigh in at about 175 pounds, Mexican wolves never get much larger than a decent-size German shepherd.

Unlike gray wolves, which can sometimes be fully black or cream-colored, Mexican wolves come tinged with a bit of rust. "There are just all these beautiful reds that come in, especially when the sunlight hits them," says Allison Greenleaf, senior wildlife biologist for the U.S. Fish and Wildlife Service's Mexican Wolf Recovery Program. "I find them to be the most beautiful wolf, even though I'm pretty biased, because I've been working on them for so long."

Mexican wolves also live in packs that are much smaller than those of their cousins, with an average of just six animals—usually a breeding pair, their offspring, and maybe a few yearlings from the previous season—per group. Meanwhile, the average pack size for gray wolves in Yellowstone National Park is nearly twice as large.

And yet, despite smaller frames and reduced pack-power, Mexican wolves still manage to punch above their weight class. For instance, it's not uncommon for a pack to take down a fully grown, 700-pound elk. They might be tiny, but they are scrappy, says Greenleaf.

They're also strategic. Wolves are highly communicative creatures, using a combination of body posture, vocalizations, and chemical scents to strategize with their pack-mates, as well as to keep rival packs at bay. While the idea of wolves howling at the moon is little more than monster movie myth, most howls do occur at night when the pack is on the hunt, suggesting that the wolf howl is also a long-distance walkie-talkie.

One piece of conventional wisdom about wolves that is actually true? The predators do tend to target the old, young, weak, or diseased. Weirdly, this means wolves may actually benefit the overall health of the populations they hunt, because when wolves cull sick individuals, there's less chance for diseases that afflict deer and elk, such as chronic wasting disease and treponeme-associated hoof disease, to spread. Some experts even contend that wolves are able to sense sicknesses in prey animals that don't look obviously ill—though if pet dogs can be trained to sniff out cancer and COVID-19, then perhaps diagnosis-by-wolf isn't so far-fetched.

What's more, with teeth evolved to tear and cut large chunks of meat and jaws strong enough to crush bone, Mexican wolves make quick work of carcasses—even ones they did not kill. This makes the predators part of nature's clean-up crew, reducing the amount of rotting flesh on the landscape, as well as all the bacteria that festers within it.

Of course, appreciation for predators hasn't really been the norm on this continent for the past few hundred years. So it's no surprise that Mexican wolves were recently hunted to the brink of extinction—a wildlife management misstep that has since required an international coalition of governments, state agencies, tribal councils, and conservation organizations to remedy. And while the effort hasn't raised Mexican wolves back to their former glory just yet, the subspecies' outlook is now more optimistic than it's been in nearly a century.

Into the Wolf Den

With just a small population of wild wolves remaining in the United States and northern Mexico, and another 360-some animals in captive breeding facilities, the total population for the subspecies remains well below a thousand individuals. Even worse? The Mexican wolves that we do have are pretty inbred. And this presents an existential problem for the future of the lobos.

"If something catastrophic happens in the environment"—say, a new disease—"it could wipe out the entire population," says Greenleaf. "So you really want that genetic variability within your population to help protect against things like that."

Thanks to a reliable captive breeding program, scientists have access to new Mexican wolf genetics every year. But releasing captive-bred animals into the wild carries a certain level of risk. After all, the animals might not know where to go or how to find food. They may also wander into areas where they come into conflict with people or even other wolves.

Fortunately, there's another way to get captive wolf genetics into the wild population.

"With cross-fostering, you're taking genetically valuable animals that are born in captivity," says Greenleaf, "and you place them into a wild wolf den to be raised by wild wolves."

Using data from GPS collars on wild wolves, scientists first look for signs of denning behavior. "When the GPS points are all clustered in one spot for several days, or maybe there are no GPS points at all, it's because [the breeding pair are] in a den," explains Greenleaf. Then, the scientists trek out into the canyons and rocky outcrops looking for a den, or a hole roughly the size of a wolf's head. Once that is located, a relay race of sorts begins, with pups being airlifted out of captivity and handed off to various teams in a race to reach the den site as quickly as possible.

If all of those steps go according to plan, then it's time for someone to go in.

"I'm like five feet two inches and 110 pounds, so I'm oftentimes the one getting sent inside," Greenleaf says. With little more than a headlamp and a cloth sack, she worms her way inside the den on her stomach. "It is a very claustrophobic feeling," she admits, "but it smells kind of neat. It's always like fresh dirt. Earthy, superclean. It's not gross at all."

Once inside, the tunnel opens up into a larger space, not unlike a little ballroom, says Greenleaf. And there, nestled in a heap, will be half a dozen furballs so fresh and new, their eyes can't even open yet. With the wild Mexican wolf pups securely in the sack, Greenleaf wiggles her way out, backward, because there isn't even room for a human to turn around.

Once outside, the wild wolf pups undergo a standard wildlife veterinary workup. This includes swabs for genetics and PIT tags for identification (just like the ones used for pet dogs and cats). Finally, it's time to introduce them to their new, captive-bred siblings.

"To get them to smell the same, we put them all in a pile and we mix them all around," says Greenleaf. "And then we'll have them pee on each other."

In the wild, animal mothers often must lick their newborns in order to stimulate urination and defecation. So the scientists simply do what the mama wolf would, but with a wet cotton ball. And with so many smells intermingled, when the wolf pack returns, they don't even notice their family has grown.

Then, the scientists back off and wait. Occasionally, they'll drop a road-killed deer nearby to give the growing pack an extra boost of nutrients—call it an insurance policy. "At the end of the day, we want them to survive to breeding age and have puppies of their own," she says. "It's a slow reward with genetics."

It can be an unpopular job bringing back a creature that can kill. And there are some who would like to see this recovery effort fail. "It's just something that happens," says Theo Guy of a recent poaching incident involving an illegally killed Mexican wolf, the details of which he could not disclose. Guy is lead field technician for the White Mountain Apache Tribe's Mexican Wolf Recovery Program.

Not so long ago, people looked to the Mexican wolf for inspiration, he says.

"Talking to our cultural expert, he has told us that way back in the old days, Apaches tried to imitate how the wolves are, how they hunted, how they were tracking animals," says Guy. "At one time there was an old Apache song that they would sing, but it was only for wartime."

These days, the songs have been replaced with the beeping sounds of tracking collars as Guy and his colleagues spend each day keeping tabs on the wolves they've already outfitted with GPS, as well as trying to catch uncollared wolves so that they, too, can be folded into the network of protection.

"Not a lot of people get to experience this," says Guy, remembering the first time he got to place his hands on a Mexican wolf. "Knowing that they are on the brink of extinction, and I'm doing a little part to help bring them back."

"You know, Mexican wolves didn't die off naturally," says Greenleaf. "They died off because people eliminated them. And so, to me, that's out of balance."

Little by little, experts like Greenleaf and Guy are building back toward a world where Mexican wolf howls aren't an uncommon sound. Even if it means transplanting one sack of urine-dabbled pups at a time.

THE CLIFF DANCER

MOUNTAIN GOAT

SPECIES: *Oreamnos americanus* **STATUS:** Least Concern **RANGE:** Northwestern North America **SIZE:** Up to 3.5 feet tall at the shoulder; weighs more than 350 pounds **LIFESPAN:** Up to 12 years

---

AT THE TOP OF THE WORLD, there is an animal built like a ballerina bulldozer. An animal with thick fur and tiny tap-dance shoes. An animal that survives extreme altitude, bitter cold, and frequent avalanches on a diet of salad, salt, and sass. An animal that suffers no fools, be they predator or paramour.

"You'd think you're looking at a polar bear," says Julie Cunningham, a wildlife biologist at Montana Fish, Wildlife & Parks, "because they're white, furry, muscular, strong, and they kind of get almost a yellowy tinge."

Friends, allow me to introduce the mountain goat. Which is not actually a goat.

"It's interesting," says Cunningham. "Bighorn sheep are very closely related to domestic sheep. But mountain goats are not super closely related to domestic goats."

If you dig into the taxonomy—which is a four-dollar word for how scientists organize stuff—you'll find that while mountain goats are in the same subfamily as true goats, known as the Caprinae, the shaggy white mountain goats technically fall within a tribe known as the Rupicaprids, or goat-antelopes. This means mountain goats are more like cousins to the true goats.

Before we go any further, perhaps it's also worth distinguishing between the mountain goat and the bighorn sheep, which is yet another member of the Caprinae, and an animal with which mountain goats are frequently confused. Mountain goats have white fur and black daggers for headgear, while bighorn sheep have brown fur and huge, inwardly curving horns the color of sandstone.

Mountain goat horns are for stabbing, but those of bighorn sheep are built for blunt force trauma.

You can also distinguish the critters behaviorally. While both are big, brawny bovids that live at high elevations, only one of them is so extreme that it earned a name with the word "mountain" in it.

"I can tell you a quote from a helicopter pilot I work with," says Cunningham. "He says, 'Sheep go where men don't go. Goats go where sheep don't go.'"

Do a quick internet search for "pictures of mountain goats in weird places" and you'll see that this is not an exaggeration. Mountain goats on sheer cliff faces? Sure. Mountain goats above the clouds? Definitely. Mountain goats bridging gaps wider than their bodies to take a little nibble of naturally occurring salt deposits? Why the heck not?

"Once I watched a goat climb to the top of a dizzying pinnacle and stand with all four feet together on a summit measuring only eight inches square," mountain goat researcher Bruce Smith wrote. "Then he raised a hind foot, scratched behind an ear, and shook the dust from his white coat, as I looked on in wonder."

Life on the Literal Edge

It makes logical sense that animals such as ringtails and spider monkeys are good at climbing. After all, if humans had built-in claws or prehensile tails, we'd be a lot more equipped for a life at height. But mountain goats? With no thumbs, grasping fingers, tails, sticky pads, or suction cups, mountain goats seem like they'd be even less able to ascend a rocky outcrop than we are.

And yet, mountain goat habitat extends from near sea level in British Columbia and Alaska all the way up to 14,000-foot peaks in Colorado. We're talking about places where the wind regularly rips at speeds of more than 60 miles an hour and temperatures plunge to minus 58 degrees Fahrenheit. So how do mountain goats not only survive, but thrive in some of the most stark and unforgiving landscapes on the continent?

For starters, Cunningham says mountain goats are built like American football linebackers. And this is especially true for males, which are known as billies, and can weigh more than 385 pounds. This size is all the more impressive considering that the ceiling of the world is not exactly an all-you-can-eat buffet fit for keeping a linebacker bulky—mountain goats survive entirely on a diet of high-altitude grasses, herbs, and other plants.

With front legs like Christmas hams, mountain goats launch themselves up, up, up and away as they scale vertical rock faces that would make most humans faint. In fact, mountain goats can leap up to 12 feet in a single bound. The power in those limbs also assists the goats as they descend down cliff faces, buffering all of the animals' weight and arresting downward motion, lest they turn into a furry projectile tumbling down the mountainside.

Of course, the real stars of the mountain goat show are the hooves. Unlike horses, which walk on a single, solid hoof (which is actually a glorified fingernail), mountain goat hooves are cloven, like deer hooves. This means that they're split into two toes, each with an enlarged fingernail that is also composed of two parts. "There is a hard edge around the hoof, but it's really soft in the middle," says Cunningham. "So they can have a hard edge to stand on, or a soft pad for grip." The two halves of mountain goat hooves can also be spread apart or splayed to provide better traction when the situation requires it.

Interestingly, mountain goats also possess dewclaws, which are kind of like smaller, side toes situated higher on their front limbs. Pet dogs and cats have dewclaws, too, and while you may have felt these while petting your pooch or tabby and concluded they are useless remnants of evolution, many species use the little nubs for extra help manipulating or gripping objects. In mountain goats, dewclaws function sort of like the brake on the back of a pair of Rollerblades. When sliding down a mountainside, the animals can flatten those forelegs in a way that puts all four hoof components into contact with the rock.

To put a finer point on just how metal the mountain goat lifestyle is, let's consider that when female goats, or nannies, are ready to give birth, they don't descend to some lush, green field where bunnies and bluebirds play. No, they go for a solo hike up onto a rocky outcrop or cliff where terrestrial predators can't follow. Which means that after a kid is born, one of the first things it must do is learn how to descend from the earth's jagged spine.

Think about what it's like trying to raise a human toddler. Every waking moment is spent preventing them from turning seemingly innocuous housewares—spoons, stairs, toilet seats—into instruments of death. Now imagine you have to do all of that next to an open balcony and the only railing is your own body.

"Mom will often orient herself below the kids, so that if a kid falls, she's there to catch them," says Cunningham. At the same time, the nanny has to keep one eye on the skies, because up here, that's where the predators come from. "Eagles will knock the kids off the cliffs," she says.

Now, this is just one of the many ways mountain goats might meet their end. Other everyday threats include starvation, avalanches, and freak accidents, because even when your body is built for life on the edge, all it takes is one bad move to bring it all crashing down.

At lower elevations, mountain goats may also fall prey to predators such as mountain lions, wolves, or grizzly and black bears. Not that mountain goats are the easiest of targets. In 2021, hikers in British Columbia's Yoho National Park stumbled onto a grizzly carcass full of holes—holes that were later determined to be the same size and shape as mountain goat horns.

Just as evolution has sculpted these animals to live hardscrabble lives on paramount and precipice, it has also equipped them to stand their own against anything they see as a threat. And that includes each other. Since mountain goats tend to jab each other in the rump instead of bashing heads, the animals have thick wads of skin on their haunches to protect them against the horns of their neighbors. And in this species, it's the female nannies who call the shots. In fact, nannies are socially dominant, to the point where males just stay away from them for most of the year. When they do approach during mating season, males abandon bravado and proceed with caution.

This is also why hikers should always give mountain goats a wide berth. Though, admittedly, when in mountain goat territory, berths can be hard to come by.

If you should be so lucky as to see a mountain goat out on the scree, Cunningham says the best thing you can do is just stay back, sit down, and enjoy watching one of the continent's most impressive mammals. You should also keep your dog on a leash and rely on your camera's zoom. "The message I'd like to convey is: Give them space," says Cunningham. "Don't approach them. Don't get all up in their business."

"They have sharp horns for a reason," she says.

THE MOUNTAIN SCREAMER

MOUNTAIN LION

SPECIES: *Puma concolor* **STATUS:** Least Concern **RANGE:** Western North America, from the Yukon Territory to Mexico; Central and South America **SIZE:** Up to 8 feet long; weighs up to 175 pounds **LIFESPAN:** Up to 12 years (in the wild)

IN CANADA, THEY'RE CALLED COUGARS. In Mexico, puma. And in much of the United States, this animal is known as the mountain lion. But with more than 40 monikers pinned to it over the years, including things like "catamount," "painter," "panther," and even "deer tiger," let's just say that *Puma concolor* is a cat of many names.

That scientific name has a meaning, too—in Latin, the word "concolor" means "single color," which for mountain lions is a tawny sort of brown. Though underneath, the cats do tend to have some lighter, creamy white coloration, as well as a bit of black eyeshadow and whisker blush, with just a dab of black on their tail tips.

My favorite name of all, though, has to be the "mountain screamer," an alias derived from the way the cats' ghostly mating calls echo off the hills and into the night, curdling many a human's blood in the process.

"Honestly, it can be really disturbing," says Tyus Williams, a carnivore ecologist at the University of California at Berkeley. "It sounds like a woman being murdered."

The mountain lion's characteristic caterwauling reveals its place in the feline family tree. For while some of the species' nicknames include words such as "lion," "tiger," and "panther," the mountain screamer is actually none of these things, since those animals all belong to the genus *Panthera,* which also includes jaguars, leopards, and snow leopards.

So who are mountain lions most closely related to? Well, it turns out they belong to the Puma lineage, which includes cheetahs and jaguarundi. (The jaguarundi is a smaller, lesser known cat in the North, but it's relatively common from Mexico through much of South America.)

Now, all of this is important because only cats in the genus *Panthera* are capable of roaring, thanks to a special ligament in their voice boxes that can stretch and vibrate. Mountain lions, along with the rest of the 35 or so feline species on this planet, lack that loud-making ligament, which means they have to make do with screams, whistles, and chirps.

And purrs. Did you know that cats that can purr cannot roar, and vice versa? As such, mountain lions are the largest purrers on the planet. Not a bad trade-off, if you ask me.

Historically, mountain lion habitat included the entirety of Mexico and the continental United States, as well as the southern portions of Canadian provinces from the Pacific Coast to the Atlantic Coast. However, as is the case with other large predators, persecution by settlers, ranchers, and farmers—not to mention the U.S. government—exterminated the cats from huge tracts of territory in the easternmost reaches of North America.

Today, catamount territory runs only as far east as Texas, save for a small, endangered population eking out a living in Florida. However, the cats can still be found in relative abundance along the continent's north-south axis, from Canada's Yukon Territory, through the western United States and Mexico, and even all the way to Argentina.

But perhaps the most interesting place to find cougars these days? That would probably be crouched beneath the Hollywood sign in the hills above Los Angeles.

It's kind of mind-boggling that a predator that sometimes weighs as much as 276 pounds and can tackle a full-grown elk could survive in close proximity to nearly 20 million people, but mountain lion sightings are on the rise in California. Heck, in 2016, one particularly famous cougar, dubbed P-22 by scientists, snuck into the Los Angeles Zoo and attacked a koala. (State wildlife officials made the difficult decision to euthanize P-22 in 2022 after a health check revealed extensive damage to the cat's head, right eye, and internal organs, likely from getting hit by a car. P-22 also had irreversible kidney damage and a parasitic skin condition—all reminders of not only what these creatures are up against while living near humans but also what they can endure.)

"Where there is prey, there is likely to be something that is eating that prey," says Williams, who is also the author of *Big Cats (A Day in the Life)*.

And when it comes to predators, few animals can hold a candle to the cougar.

The Mountain Lion Giveth

With keen vision and camouflage that helps them blend into the scrub, mountain lions stalk their prey, sometimes for longer than an hour in pursuit of a single target. Targets can include anything from mice and porcupines to moose and wild horses. The cats can leap nearly 10 feet straight up into the air, and they use this powerful pounce to launch themselves at unsuspecting prey animals, which they then sink their retractable claws into while biting at or around the neck—either to crush the trachea and suffocate larger animals or, sometimes, sever the spinal cord from behind.

"They will also hide in trees," says Williams, "which is actually really cool, because they will literally jump down from a tree on top of their prey." Death from above.

Of course, while a mountain screamer could totally kill you, in all of North America, just 17 people have lost their lives to these animals since the 1970s. But even with these extremely low odds, you can further reduce the chances of a nasty encounter by hiking and camping in groups and making noise as you trail-run or trek. "I'm not saying you need to run through the forest screaming 'Bohemian Rhapsody,'" says Williams. And in fact, please don't. "But even allowing your keys to dangle off your backpack lets animals know that you're coming."

If you somehow still have a cougar staring at you from the middle distance, then you should consider that the cat thinks it can take you. In which case, Williams says,

"Don't act like prey." Stand tall, wave your arms, throw rocks and sticks, and yell that cat down like it spoiled the end of a new movie. All the while, you should be slowly backing away. As a last resort, if the cougar comes at you, fight like your life depends on it—because it probably does.

Chilling as it might be to stare into the amber eyes of one of these apex predators, the cats give back far more than they take. For starters, some studies have shown that mountain lions keep prey populations, such as mule deer and bighorn sheep, in check.

But the cats don't actually even have to kill grazers in order to influence their behavior. This is because a predator's mere presence can influence prey animals to act differently, even when the predator is just lurking nearby rather than hunting. Because predators make prey more vigilant and more flighty, they can reduce how much those animals graze an area, making them more likely to nibble and move rather than mowing an area down to the nubs. And this effect can radiate outward in surprising and impressive ways.

Fear-driven prey distribute seeds farther, says Williams, leading to more diverse ecosystems. And less grazing in waterways means more roots to hold back soil, reducing erosion, which in turn triggers changes in the creek or river's water quality, which—bing, bang, boom—makes the area more habitable for fish, insects, and the birds that eat them.

And it's not just prey. Cougars change the behavior of other predators, too, because the threat posed by one predator to another alters the way everyone moves through a given ecosystem, as well as what foods they eat. "It's called niche partitioning," says Williams, "and it's the way that animals separate themselves by time, space, and diet."

At the same time, mountain lions create islands of biodiversity everywhere they leave a kill. One study conducted in the Greater Yellowstone Ecosystem found that cougar kill sites provided nourishment for 39 species of mammals and birds. Another study revealed that carcasses left by the cats supported five times as many beetles as sites without a carcass. And when they sat down and tried to identify all those scuttling insects, the scientists were gobsmacked to find 215 different beetle species—a three-ring circus of life—all brought together as a simple result of mountain lions doing what mountain lions do. All of which led the experts to conclude that the cats qualified as ecosystem engineers, or animals that modify their habitats in meaningful ways, like beavers and humans.

Add all of that up and you'll understand why many experts consider mountain lions to be a keystone species, or one on which the ecosystem depends. Still others label these cats as a bellwether species, because when their population is healthy, you know that the ecosystem itself is thriving. Conservationists also call the cougar an umbrella species, since one cat can require 13 times more area than a black bear or 40 times as much area as a bobcat to make a living.

As I said in the beginning, you can call a mountain lion many different names. But maybe there's room for one more:

Absolute legend.

THE IRON-TOOTHED LUMBERJACK

NORTH AMERICAN BEAVER

SPECIES: *Castor canadensis* **STATUS:** Least Concern **RANGE:** Most of North America **SIZE:** Up to 4 feet long; weighs up to 65 pounds **LIFESPAN:** 10 to 12 years

---

BEAVERS ARE RODENTS, and rodents of unusual size at that. The heaviest beaver on record tipped the scales at a whopping 110 pounds, but even at an average weight of 60 pounds, beavers are the largest rodent in North America. Worldwide, only South America's capybaras grow larger. And like their giant cousins to the south, evolution has sculpted the mighty beaver to be a superhero of the semiaquatic realm.

You no doubt know how beavers swim and build dams and lodges upon waterways. But did you know beavers have an extra eyelid that is see-through and protects their eyeballs like goggles while they're underwater? It's called a nictitating membrane. Beavers also have a second set of lips behind their teeth that acts like a waterproof valve and allows them to carry branches and rocks in their mouths without muck sloshing down their throats.

By the way, those teeth? "They're iron-reinforced," says Ben Goldfarb, science writer and author of *Eager: The Surprising, Secret Life of Beavers and Why They Matter*. That's why beaver chompers are a rusty orange rather than pearly white. Beavers use those metal-reinforced munchers to tear into the sides of trees to expose their favorite food: cambium, or the growing part of the tree's trunk. Tree meat, basically.

Then there's the beaver tail, which is essentially a Swiss Army knife attached to the rump. Beaver tails are huge, black, scaly paddles that act like rudders in the water and kickstands on land. A beaver's tail is also its alarm system, says Goldfarb, because when danger is near a beaver will thwack its tail against the water's surface so hard, it sounds like a gunshot. The sound startles anything nearby and tells other beavers to dive below the water, where they can remain for up to 15 minutes.

The beaver's tail has other useful purposes. It's a larder and an air conditioner. When beavers have plenty of food, extra energy gets stored in the tail in the form of fat, which the beaver can then draw upon over the winter when high-quality foods are scarce or it's too dangerous to venture outdoors. On the flip side, in the summer a beaver's tail can help shed heat, as the things are loaded with thermoregulatory blood vessels.

In cartoons, you may have seen beavers using their tails to slather mud upon their dams, but that's actually one thing beavers don't use their multipurpose tails for.

Now, I'm going to guess there's at least one other thing you probably don't know about beavers. And that's the fact that in August of 1948, they flew through the skies above Idaho.

Flying Beavers

In the days before European colonization, scientists estimate that North America held as many as 400 million beavers, or more beavers than the current combined human populations of the United States and Canada. From the watersheds of the Colorado River and Rio Grande in the south to the uppermost reaches of Alaska and Canada, these tree-gnawing rodents ruled the continent's waterways, if not with an iron fist, then at least with an iron tooth. That is, until some fashionistas figured out that beavers didn't just make dams, they made dang good headgear.

"Beavers have two layers of fur," says Goldfarb. "Long outer guard hairs and soft underfur, which trappers called the beaver wool."

With this combination of fur types, beavers basically come equipped with "armor, life preserver, and dry suit," writes Goldfarb in his book. A single square inch of beaver pelt boasts 126,000 hairs, more than a human has on their entire noggin. And because beaver underfur has tiny hooks that allow it to lock together like Velcro, hatmakers could use it to craft all sorts of chic designs, such as the Regent, the Wellington, and the Paris Beau. As you can probably tell from the names, these hats were designed for high-society customers, not backwoods frontiersmen, and this led to an unfortunate outcome.

Rather quickly, beaver hats became a status symbol, causing a spike in demand. Then, as fur trappers raced to keep supplying that demand, beaver populations tanked. And so, in just a few hundred years, the free market managed to annihilate a once-bounteous rodent resource.

"By 1850 or so, it would have been almost impossible to find a beaver in the United States," says Goldfarb. "We went from likely several hundred million beavers to having these animals basically functionally extinct in a really short period of time."

Now, what happens next is actually pretty incredible. Taking note of what had unfolded, a whole bunch of government officials, conservationists, and former trappers banded together to bring the bucktooths back from the brink.

Beginning in 1901, New York state started reintroducing beavers, which quickly multiplied, as rodents do. Soon, there were so many beavers that they started recolonizing nearby Massachusetts. Later, New York shared its beaver wealth with Vermont, jump-starting populations in the Green Mountain State.

Then beaver fever spread out west, where the U.S. government released around 600 of the semiaquatic rodents into California, Wyoming, Oregon, and Utah. Then North Dakota and Washington. "From the marshes of Maryland to the coasts of Oregon, the beaver tide began to rise," writes Goldfarb.

All of which brings us back to Idaho in 1948. There, too, beavers were on the rebound. But the rodents had the audacity to rally in places where humans didn't want them, like around their houses and farms, which the beavers kept flooding with their dams. At the same time, Idaho was experiencing hydrological issues upstream—water moving too quickly off the land, leading to problems with erosion and silt. Problems that could be solved by lots of little dams.

"At first they tried moving the beavers on horseback," says Goldfarb. "But the beavers didn't like that, and the horses definitely didn't love having big, smelly rodents strapped to their backs."

Solution? Someone got the bright idea to drop beavers out of the sky.

Only the gods know what the giant rodents must have been thinking when the humans who loaded them into wooden boxes—boxes with air holes drilled into them and WWII-issue parachutes attached—kicked those crates out of a small airplane some 500 to 800 feet above the ground. Nor do we know what sort of fear one particular beaver might have experienced when a small opening in its box allowed it to squirm free and climb atop its rapidly descending vessel. The critter must have stared in wide-eyed wonder at the unknown wilderness below—mountain streams, boulders, stands of aspen and willow—rising up toward the unfortunate beaver at a mighty alarming clip. All we do know is that when this parachuting rodent got within 75 feet of the ground—roughly the height of a six-story building—it either slipped or jumped off the box in a skydive that it did not survive.

Despite the tragic fate of this single beaver, the rest of the mission went swimmingly, with beavers dropped into the Idaho wilderness in male-female pairs, like a sci-fi plot to repopulate Earth after the apocalypse.

A year layer, when conservation officers returned to what is now known as the Frank Church–River of No Return Wilderness Area, they were amazed to discover the beavers' touch had transformed the area into a quilt of ponds and wetlands. Why? Because beavers are what's known as ecosystem engineers, or critters that can manipulate their environment.

When beavers dam up small creeks and turn them into giant ponds and sprawling wetlands, they're doing it to protect themselves. "A beaver out on the land is very vulnerable to predation," says Goldfarb. But doing so also creates habitat for everything from moose (which eat aquatic plants) to muskrats, frogs, and salamanders. Dragging woody materials into the water creates homes for insects, which fish eat. When beavers are around, moth biodiversity blossoms. Beaver lodges provide shelter for mink, river otters, fish, and birds such as great blue herons and wood ducks.

And that's not even to mention the array of ecosystem services beaver-made wetlands provide for humans. "Their ponds filter out pollution and act as a firebreak slowing the spread of destructive wildfire. They store water in the face of drought, and they're building thousands of little reservoirs for us up in the high country, keeping water on the landscape," says Goldfarb. "They're sequestering carbon. They're mitigating big flood events."

According to one estimate from the Escalante River Basin in Utah, reintroducing beavers to a single watershed is worth between $219 million and $1.4 billion in projects we would otherwise have to do ourselves. Jobs that beavers happily do for free.

And that, frankly, is more than enough to make a beaver believer out of me.

LICENSED TO QUILL

NORTH AMERICAN PORCUPINE

SPECIES: *Erethizon dorsatum* **STATUS:** Least Concern **RANGE:** From northern Mexico to Alaska and east to Newfoundland **SIZE:** Up to 3 feet long; weighs up to 20 pounds **LIFESPAN:** Up to 18 years

---

THE NORTH AMERICAN PORCUPINE weighs about as much as a sledgehammer, spends most of its life straddling tree limbs, and is covered in a weaponized hairdo that's killed many a mountain lion, pet dog, and even one extremely unfortunate woodsman.

Porcupines are rodents—pretty big ones, in fact. At around 20 pounds, porcupines are actually the second largest rodents in North America. Only beavers top them.

Of course, the one thing every person on Earth knows about porcupines is that the animals are covered in tens of thousands of dagger-sharp quills. In fact, the North American porcupine's genus—*Erethizon*—comes from the Greek word for "irritable," which is what you'll be if you ever meet a quill up close.

But despite this reputation, porcupines are actually kind of sweet. "They're not really an aggressive animal at all. They're a defensive animal," explains Jessy Coltrane, a wildlife biologist with Montana Fish, Wildlife & Parks.

When she was completing her Ph.D. in Alaska, Coltrane kept tabs on two porcupine populations, some of them wild and some captive, as she sought to better understand how the spiky ones survive on a nutrient-poor diet of pine needles, tree bark, and twigs. She got to know lots of porcupines in the process. And while occasionally there'd be one that was standoffish, mostly, the porcupines acted like puppy dogs.

"I'd say within a week, nine out of 10 of them were handleable. They wanted to eat out of your hand or crawl up your leg like you were a tree," she says.

Of course, this gentleness does not apply to predators, and, believe it or not, some animals do actually try to attack porcupines. A mountain lion can climb below a porcupine in a tree and bat it to the ground, whereupon the cat pounces and kills the unlucky porky. Lynx, bobcats, coyotes, wolves, wolverines, and great horned owls will also take a run at a porcupine on occasion, but these occurrences are thought to be on the rarer side.

But far and away the biggest predator of porcupines wherever it still exists is a little, lesser known member of the Mustelid family called the fisher. Fishers weigh as much as a toy poodle, but look like tiny, long-tailed grizzly bears. Using a combination of speed and cleverness, the fisher dances a do-si-do around its prey until the porcupine trips up or turns too slowly. Then it strikes, fast and hard, at the porcupine's tender nose, mouth, and eyes. Circle, lunge, strike. Circle, lunge, strike. Again and again, until the porcupine goes blind or becomes disoriented enough to let down its guard, at which point the fisher flips the porcupine over and starts to disembowel it like a shucked oyster.

The Secret Life of Quills

While porcupines cannot throw, shoot, launch, or fire their quills into the air like you've no doubt seen in cartoons, these structures are incredibly impressive in their own right. In fact, having quills is kind of like if 30,000 Swiss Army knives just up and sprouted out of your skin.

A quill is a special kind of hair, and each one grows out of a follicle in the porcupine's skin. When a porcupine loses a quill, its body begins manufacturing a new one, a process that can take between two and nine months.

Unlike more typical hairs, quills have been modified by evolution to be longer, sharper, and stiffer. Each quill is coated in extra layers of keratin, the protein other animals use to build hooves and horns (keratin is also the basis for human hair and fingernails).

Quill length differs by the part of the body it protects, but the longest spines on a North American porcupine usually top out at about four inches. In fact, when the species grows longer, thicker fur to weather the bitter cold of winter, its quills might look like they've disappeared under the fluff. They are still present though, ready to strike at a moment's notice.

Look closely, and you will find that each quill is coated in overlapping scales—like the belly of a snake or a carefully constructed dish of au gratin potatoes. Rubbing a fingernail from the point to the base, the quill feels smooth, but travel the other direction and your nail will catch slightly, as though it's being dragged across an extra-super-fine-grit sandpaper. And this arrangement serves a diabolical, underappreciated purpose.

When a porcupine defends itself, the quill's sleekness enables it to penetrate the skin of an attacker easily. But try and pull the same quill back out and the microscopic barbs that encircle the spine hold it fast as a fish hook.

Fascinatingly, quills can prevent an attack without ever even touching a predator. This is because North American porcupine quills are black at the base and white at the tips—colors that show up well in the night vision of most nocturnal animals. This is what's known as aposematic coloration, or warning colors. And the porcupine enhances this effect with long guard hairs all over its body that are also tipped in white. These tendrils look like the quills hidden below the fur underneath, but they are completely harmless.

To illustrate the point, both the white sections of a porcupine's quills and the white stripes of a skunk glow when exposed to ultraviolet light, whereas the fur of your white Jack Russell terrier would not. This is because domesticated dogs have no reason to advertise themselves to nocturnal animals, while nocturnal wanderers, like porcupines and skunks, have everything to gain from warning a predator away before it causes damage. In other words, evolution hasn't just given these animals white coloration, it's added brightener compounds that make them impossible to overlook.

Because quills are built out of a spongy matrix with air inside, they also help insulate porcupines against the cold as well as serve as a sort of built-in life raft when the rodents go for a swim. Each quill is also coated in an array of free fatty acids shown to kill gram-positive bacteria, such as the microbes that cause gangrene and tetanus. And this comes in handy when male porcupines battle over a female or when the animals impale themselves on their own quills after falling out of a tree, which apparently happens. So even if a porcupine winds up with needles in its skin, the built-in antibiotics prevent those wounds from festering.

Of course, when you study porcupines, getting quilled is just part of life. "I mean, I used to get quilled all the time," says Coltrane without a hint of regret.

Even welder's gloves, which are built to withstand white-hot molten metal upwards of 2,500 degrees Fahrenheit, were not enough to stop a quill from once driving straight through the meat of a colleague's index finger. Coltrane performed the extraction herself by pulling the quill clean through.

If you don't remove an embedded quill immediately—and most animals cannot—the quill will start to shimmy itself deeper every time the afflicted animal moves. Every snarl, every yawn, every lick stretches skin and contracts muscles in a way that brings the quill to life. Over time, the whole spine can disappear below the skin and move through muscle, lodge itself in joints, and in extreme cases, even puncture organs, causing sepsis and death. This is also why Coltrane was so concerned when the same colleague who had his finger impaled another time got the tip of a quill nestled up right against his eyeball. (She performed that extraction, too, but only after the local eye clinic refused to touch him. Porcupine scientists go hard.)

Researchers have documented many instances of death-by-porcupine, and scores more instances surely go unnoticed out in the wild. In 2015, a mountain lion died after porcupine quills lodged in her chest and migrated into her lungs. A review of interactions between porcupines and birds, also from 2015, found numerous fatalities, including a raven that died with quills in its gizzard and heart, a deceased golden eagle riddled with quills in its intestines, and even a bald eagle that died with a quill lodged in its throat. The birds most likely ingested the quills while scavenging from a porcupine carcass.

There are even a few examples of porcupine quills turning up in the flesh of fish, such as bull trout and rainbow trout. While no one has seen it happen, scientists' best guess is that the large, predatory fishies tried to attack a porcupine while it was swimming and wound up with a face full of nature's harpoons instead.

And it's not just wild animals. In 1934, a 30-year-old woodchopper from New Hampshire ate a porcupine-meat sandwich, resulting in a quill getting lodged in his stomach wall. "The course was stormy," according to a report of the surgery to remove the barb. Eight days later, the man died.

Amazingly, 20 years and a few days later, the curse of the porcupine sandwich played out again, this time in Vermont. A 51-year-old woodsman had shot a porcupine and then cut off its head with an ax so that he could cash in the bounty. (Historically, many states have at one time or another offered cash rewards for proof of killing porcupines.) The man was smart enough to wear leather gloves while handling the animal but not smart enough to keep his gloves away from his lunch. Yes, a quill found its way into his sandwich, which found its way into his bowels. This time, the woodsman lived, but the moral of the story remains ...

Don't mess with porcupines.

THE APEX PREDATOR

ORCA

SPECIES: *Orcinus orca* **STATUS:** Data Deficient **RANGE:** Worldwide; in North America, western coast, eastern Arctic, and Gulf of Mexico **SIZE:** 30 feet long; weighs up to 11 tons **LIFESPAN:** Up to 90 years

---

USUALLY JET-BLACK, with white eyespots and underbellies and a grayish patch called a saddle just behind their dorsal fins, orcas look a bit like the glistening pandas of the sea—you know, if pandas were cunning pack predators that were weighed in tonnage.

The largest orca ever recorded, a male, clocked in at a walloping 32 feet long and 22,000 pounds. That's about as long as a three-story building is tall and as heavy as two African elephants. But there might be an even better way to put this massive marine mammal's size into perspective.

"When they hit puberty, the male's dorsal fin can get to be four and a half to six feet tall," says Alisa Schulman-Janiger, co-founder of the California Killer Whale Project. In other words, they have a fin on top that's as big as *you*.

Put another way, orcas can grow more than 10 feet longer than great white sharks. And not only do they weigh more than a great white, they weigh more than a great white sitting on top of a full-grown *Tyrannosaurus rex*.

Unlike some of the animals in this book, orcas, or killer whales, are a household name—even if it's a confusing one. For starters, we often call them by the common name "killer whales," though the species is actually the largest member of the oceanic dolphin family (known as the Delphinidae). However, because the Delphinidae are just one branch within the greater toothed whale family, known as the Odontoceti, it means that killer whales are actually whales, too. Put another way, killer whales are dolphins that are whales. Or less confusingly, killer whales are both whales and dolphins.

What's more, there's been a push in recent years to use the species name "orca" when referring to these animals, because some think that "killer whale" paints the creatures in a rather unsavory light. But plenty of whale experts disagree with that logic. "They are the killers of whales," says Schulman-Janiger, who has been studying the species since the 1970s.

Many human cultures have hinted at the killer whale's carnivory when choosing a name. In Portugal, the animal goes by *baleia assassina,* or "assassin whale." Similarly, in Aleut, *polossatik* means "the feared one." And you only need to try to pronounce the German word *mörderwal* out loud to realize its meaning: "murder whale."

In any event, Schulman-Janiger uses both "killer whale" and "orca" to describe the species, and you can, too. "They're interchangeable," she says.

With a presence in every one of the world's oceans, and a range greater than any other whale or dolphin, some consider killer whales the most widely distributed mammal on Earth—aside from us, that is. But here's the thing about that far-flung distribution: While scientists currently recognize just one species of orca, if you could hop across the globe and spy on these animals, you would quickly learn that each population differs from the next.

"A killer whale is not just a killer whale," says Schulman-Janiger. "It depends on where it is, and what kind."

Variety Is the Spice of Life

On North America's western coast, scientists recognize three main killer whale ecotypes, or populations that have specialized to local conditions. First, you have the resident killer whales, which tend to keep small home ranges centered on large fish populations that they hunt. On the other hand, transient killer whales, also known as Bigg's orcas, like to roam around and prey upon marine mammals. Finally, offshore killer whales have the largest home ranges of all and, as such, are the least understood. Scientists think offshore orcas specialize in fish though, because their teeth are often ground down to nubs—likely as a result of gnawing on tough-skinned fishes, such as sharks.

Importantly, each of these orca varietals might differ from the others in color pattern, fin shape, body size, diet, hunting strategy, call structure, and even genetics. They do not interact, nor do they interbreed.

And many more ecotypes exist. Around Antarctica, pack ice orcas (also known as Type B's) famously create waves to knock seals off sea ice, while Type A's hunt migrating minke whales and Type C's go after Antarctic toothfish. Each of these ecotypes looks completely different from the whales native to North America because their skin blooms with algae that lightens their black coloration to gray and gives their whites a dash of yellow.

Scientists also have their eyes on several more populations likely developing their own ecotypes, such as the Strait of Gibraltar orcas that eat tuna, so-called tropical orcas in Hawaii and the Gulf of Mexico, and a very small resident group in New Zealand that weirdly seems to go after both fish and mammals.

Variation also exists within these ecotypes. For example, within the residents there are Northern, Southern, and Alaskan communities. The first two eat salmon, almost to exclusion, while the Alaskan residents are more generalist in their fishy tastes. But even when fish are scarce, none of the residents will switch to eating marine mammals. Almost as if, to them, marine mammals are taboo in the same way that Americans and Canadians don't often eat insects, but fried grasshoppers—known as *chapulines*—are a rather common snack in regions of Mexico. (For what it's worth, humans also differ regionally in skin color, body size, language, hunting or farming technique, and numerous other ways that are not all that dissimilar to orca ecotypes.)

"Their whole tradition, their genetic makeup, everything pushes what they've been taught by Mom and their mom's mom," says Schulman-Janiger. "Their cultural traditions are very strong."

Interestingly, some other animals appear to be able to distinguish between orca ecotypes, too. Because sometimes scientists see seals or gray whales hanging out with orcas—but not the orca ecotypes that hunt them.

However, there is also flexibility within all that variety.

In the last hundred years or so, orcas have started to invade Canada's Hudson Bay, plunging deeper into waters once blocked by nearly eternal sea ice. (Remember that giant dorsal fin? It actually prevents orcas from cruising through too much surface-level ice.) Climate change opened the door to these waters, and orcas came waltzing in. And that's bad news for the Arctic's other marine mammals.

"People have seen killer whales throwing narwhals and belugas into the air," says Kristin Westdal, science director for Oceans North, a charitable organization that supports marine conservation in partnership with Indigenous and coastal communities. "They're really known to play with their food."

While incursions into the eastern Arctic may eventually become a cemented tradition in that population's culture, other behaviors are more like fads. In one example from 1987, a female Southern resident orca native to Puget Sound killed a salmon and then swam up to the water's surface with the dead fish slung across her head. "And then another one did it, and another one, and another one," says Schulman-Janiger. "It became a thing where you'd see multiple whales get a salmon and stick it on their head. It was called the salmon hat." Like a hit song or dance, the behavior even spread to neighboring pods before mostly disappearing altogether. (In 2004, orcas in the northwest Pacific briefly revived the trend; perhaps salmon hats are coming back into style.)

More recently, fish-eating killer whales off the coast of Portugal have begun disabling yachts by attacking their rudders. And while there have been no injuries to humans, the fact that this behavior has been documented more than 500 times between 2020 and 2023 has led to lots of questions about whether orcas might be rising up against the humans.

Nah, probably just another fad, says Schulman-Janiger.

Interestingly, there's also historical evidence that the Thaua people and orcas actually used to team up to hunt other whale species collaboratively in the waters off Eden, Australia. Unfortunately, scientists believe that population of killer whales died off in the early 1900s.

Of course, all of this is just a smidgen of what we know about these highly intelligent animals and their complex cultures. Orcas actively communicate with each other as they hunt. They strategize and problem-solve. Some scientists even consider the Type B's' wave-washing technique a form of tool use, since the animals manipulate water to perform a task they couldn't do alone. Also, unlike nearly every other mammal on Earth, killer whale females outlive menopause, which means old granny orcas are still out there swimming around well past their ability to reproduce. And scientists hypothesize that this strategy has evolved to allow orca societies, which are female-led, to benefit from all that knowledge accumulated by the older generation. Literally, orca grandchildren have a better chance of surviving to adulthood when their pod still has its grandma.

As highly emotional animals ourselves, people seem especially interested in the mounting evidence that orcas can grieve. In one case from 2018, a 20-year-old Southern resident orca mother nicknamed Tahlequah carried the body of her recently deceased calf for 17 gut-wrenching days as the human world looked on. Scientists still aren't sure why, but they do know that another member of this same pod performed a similar behavior more than 15 years prior.

"Even though we have a lot of questions that are starting to be answered, there are going to be ones that will never be answered. That's part of the mystique of these animals," says Shulman-Janiger. "They are aliens on this planet. Their world is the ocean. And we can only see a very small percentage of it."

THE NIGHT NINJA

RINGTAIL

SPECIES: *Bassariscus astutus* **STATUS:** Least Concern **RANGE:** From Oregon in the U.S. to the Mexican states of Guerrero, Oaxaca, and Veracruz **SIZE:** Up to 2 feet long; weighs up to 2 pounds **LIFESPAN:** Up to 9 years

---

FACE OF A CAT? Check. Ears of a bat? Check. Body of a ferret? Check. Tail of a lemur? Check, but with a dash of feather duster. Add it all up, and you get a criminally under-known mammal called the ringtail.

"They really are just ridiculously adorable," says Lindsay Somers, a biologist for the Oregon Department of Fish and Wildlife. "The small nose, the cute little paws, the fuzzy faces, the big eyes. It's a combination of everything."

Not surprisingly, this animal has been called many different names over the years, including miner's cat, coon cat, raccoon fox, cacomistle, bassarisk, and civet cat. Even its scientific name, *Bassariscus astutus,* is a misnomer; it means "clever little fox."

In truth, this lithe and linear mammal is neither fox, nor cat, nor civet. With a ringed tail and slight hint of a bandit mask, the ringtail is actually a member of the Procyonidae family—a curious branch of mammals that includes characters such as the coati, kinkajou, olingo, and olinguito, as well as everybody's favorite rascal, the raccoon.

Huge, forward-facing eyes, black and glistening like olives, help ringtails navigate their nocturnal realm. On their muzzles, long whiskers—technically called mystacial vibrissae—relay information about the world around them and enable a ringtail to slink through narrow cracks, crevices, and tunnels without a flicker of light. The underside of each front paw is also equipped with long, touch-sensitive hairs that allow the ringtail to feel what it can't see, especially when it's digging around in a rock crevice in search of a plump woodrat.

Unlike the ring-tailed lemur, which tends to hold its black-and-white tail high in the air as it walks on all fours, the ringtail usually holds its feather boa–like tail low and straight. The one exception seems to be when the animal has to cross a large, open area where it might be more exposed to predators, such as the great horned owl, bobcat, or coyote. In this case, the ringtail will actually arch its tail over its back, perhaps in an attempt to make its bunny-size frame appear larger and more imposing.

The animal's tail also helps with balance while performing its nighttime ninja maneuvers. While ringtails inhabit a variety of ecosystems across their semiarid range, they need to climb no matter where they live. Ringtails have semiretractable claws that allow them to walk up trees and cacti like flies on a wall. On the way back down, the ringtail can rotate its hind feet 180 degrees—a trait shared by raccoons and other members of the Procyonidae family. Twisting the hind paws and claws around provides a secure grip and allows the mammals to descend headfirst with ease, instead of scooching their way down backward like a cat.

Boulders and ledges make up the ringtail's own personal parkour course. For short gaps, a ringtail activates its power leap. Chimneys, or cracks large enough to crawl inside, present an unpassable barrier to many creatures, but the ringtail knows how to stem—wedging itself in the space with its back against one wall and its feet against another, and then waltzing right up or down as need be. The ringtail can also conquer wider chasms through a technique called ricocheting, where it bounces from surface to surface doing its best impression of a fuzzy bullet, the black-and-white tail following closely behind like motion blur in a photograph.

When confronted with a situation it cannot escape by other means, the ringtail first lets out a screech like a Yorkie caught underfoot. If that doesn't scare the bejeezus out of whatever is bothering it, the ringtail will then lift its tail and let a pair of anal glands do the talking. And trust me, you don't want to hear what they have to say.

Unlike the better known skunk, ringtails cannot squirt their liquid stank into the eyes or face of an assailant. But don't discount their ability to clear a canyon with their fumes. The milky white secretion ringtails produce is gamy and ripe in all the worst ways.

"It's pretty sweet, though I wouldn't call it skunky," says Somers, who scrunches up her face recalling a research project that necessitated she load a ringtail into her truck while regrettably forgetting to put down a towel first. Some of the animal's juices got onto the seat and lingered in her cab for weeks. "Great memories," she says, almost gagging just thinking about it.

Of course, grossing out predators and scientists is just one use ringtails have for their secretions. Because the animals regularly mark their territory in captivity, scientists suspect that they're also tagging the natural world in secret scent markings.

Ringtails are fastidious toilet-users, teaching their pups at a young age that you're not allowed to just go potty wherever you want. Rather, the species makes use of dedicated latrines, or bathrooms, where they go number two. Curiously, this might have less to do with cleanliness than with communication. Rhinos, wombats, badgers, hyenas, and tiger quolls all mark their territory with the chemicals stowed away inside their poop, which they pile up in great quantities. Ringtails, too, may be part of the poo crew.

Ringtails live mostly solitary lives, unless they're meeting up to breed or raising young. They can also be itinerant wayfarers, with home ranges spanning anywhere from 20 to 1,700 acres depending on sex and habitat, says Somers. Some researchers have noted that ringtails tend not to sleep in the same spot for more than three days in a row. And new mothers sometimes move their babies as soon as 10 days after birth. "You can only eat so many mice in one region," says Somers.

While the ringtail is the smallest member of the raccoon family, it is also the most meat-loving. Ringtails have been known to hunt down everything from pack rats, mice, and rabbits to bats, snakes, lizards, fish, birds, insects, and amphibians. They also enjoy prickly pears, juniper berries, and persimmons.

In any event, it would stand to reason that these creatures have evolved sophisticated methods for avoiding each other during walkabouts, while also still being able to locate their own kind now and again to reproduce. It's also possible that this all happens through vocal communication, since the species is known to emit an array of loud, piercing chirps and barks. However, at the time of this writing, many aspects of the ringtail's after-dark lifestyle remain a mystery.

The Miner's Cat

In the years following the 1849 California Gold Rush, mining towns started popping up all over the American West. Prospectors flocked to such places in search of not just gold but also silver, zinc, copper, and lead. And as is often the case, where the people went, the critters followed. But not the ringtails—at least, not at first. They were still out in the canyons stalking toads and doing battle with owls.

But then a funny thing happened. In time, the predators followed the prey, and it became nothing to see a ringtail slinking between shanties or crouched on top of a saloon roof. Indeed, the ringtail seems to have been just as common and accepted as an alley cat or stray dog.

It didn't take long for the prospectors to realize that the ringtails were actually providing them with free rodent-removal services. Some even invited the predators into their homes by setting up nest boxes next to their woodstoves. Small holes gave the ringtail access to the artificial den, which was warmed by the fire, and at night, the raccoon cousins paid their rent by devouring any vermin that dared show their faces. This is also how the ringtail earned another of its common names: the miner's cat.

These days, you probably wouldn't want a ringtail living in your home, even if they're great mousers and cute to boot! Scientists have since learned that the species can carry rabies, canine distemper, and a whole litany of parasites, including fleas, mites, lice, ticks, and nematodes. Some of which can carry diseases of their own, and none of which you want beside your bed. Also, while that scent-marking and poop-piling behavior plays a necessary role in the ringtail's existence in the wild, such habits are unlikely to be popular on the home front.

If you live in ringtail country and start to hear scratching up above at night, you might wait a few days and see if the ringtail moves on. Once the animal has left, you can hopefully find and then fill in or patch whatever entry point it was using to get into your home. However, if the prey is plentiful or the ringtail is about to give birth, the intruder may just decide to take up permanent residence. At that point, your best bet would be to call a local professional who can safely trap and legally relocate the ringtail.

While the ringtail's secretive nature makes population estimates difficult, it's clear that the tiny, agile predators are important assets to their ecosystems. They mow down smaller critters, which themselves can carry disease and cause problems for people. Chances are good that most people on this continent will never even catch a glimpse of the fluffy-tailed phantoms, but it's cool to know they're out there, ricocheting through the night.

THE SPEED EATER

STAR-NOSED MOLE

SPECIES: *Condylura cristata* **STATUS:** Least Concern **RANGE:** From Florida to Manitoba
SIZE: Up to 7.5 inches long; weighs up to 1.9 ounces **LIFESPAN:** Thought to be 3 to 4 years

---

LOTS OF PEOPLE THINK MOLES ARE BLIND or lack eyes altogether. But this is false. All moles have eyes—though some species are probably near-sighted and color-blind.

One thing moles do lack is external ears, or what scientists call pinnae. This makes sense, though. In the subterranean realm, open orifices can get clogged with dirt, so evolution did away with the moles' ear holes long ago. Despite the loss, most mole species can still hear rather well by detecting vibrations and low frequencies rumbling through the soil.

Seven species of moles inhabit North America. All of them have their own cool stuff going on, but we're going to focus on star-nosed moles because, well, they're the moley-est moles of them all.

Star-nosed moles are covered in soft, silvery fur, except for their giant paws, which are meaty, mostly furless, and the size of the mole's head. Truly, there's no sugarcoating these front appendages, which come with rake-like claws and look for all the world like the gigantic, rubbery hands that might go with a werewolf costume.

Of course, these oversize atrocities have a purpose—they help the tiny mammals tear their way through soft soils in search of prey.

Unlike other mole species that dig tunnels through the dirt, star-nosed moles are semiaquatic. Their habitats include swamps, glens, and wetlands, where they burrow through muck. (Star-nosed moles are found from Newfoundland and Labrador in Canada all the way over to Manitoba and U.S. states surrounding the Great Lakes, as well as south along the Appalachian Mountains and in coastal pockets as far south as northern Florida.) And while you'll likely never cross paths with one in the wild, you'd be almost as likely to do so at the bottom of a stream as you would in the ground, because star-nosed moles are powerful underwater predators, even in the dead of winter.

"Star-nosed moles don't hibernate," says Kenneth Catania, a neuroecologist at Vanderbilt University. "And they seem to dive more frequently in the winter than other times." In fact, from what he's seen, Catania says you can think about star-nosed moles "like miniature penguins." But instead of eating fish and krill, these critters spend several months diving below the ice and vacuuming up all the insect eggs and larvae that thought they'd found a safe place to spend the winter.

Look closely at a star-nosed mole, and you'll also see a pair of beady, black eyes on either side of its cone-shaped face, a short tail, and a mouth full of jagged teeth. But honestly, there's no way you're going to be bothered worrying about any of these things, because there's one part of the star-nosed mole's anatomy that demands our undivided attention.

Yes, friends, it's time to talk about *the star*—a sensory organ unlike any other known to science and a structure that's earned this mole a place in the wildlife record books.

"The star is such a bizarre adaptation," says Catania. "It's really as unusual a feature as you can imagine for any mammal. For any animal."

Stars on the Brain

If you've never seen one in person, the star-nosed mole's fleshy protuberance looks like what might happen if you shoved a stick of dynamite up an elephant's trunk. It's pink, it's splayed, and while the word "star" implies that the structure is composed of five or six lobes, in truth, the star-nosed mole's schnoz sports 22 pink, worm-like tubes. Scientists call them nasal rays.

Scientists have been fascinated by this structure for decades. And while several theories about its purpose have emerged, Catania is one of the few researchers around to actually test them. Some speculated the star is for grabbing or clutching things in the soil, but while the star is full of tendons, which allow the structure to slosh around like a mop head, it does not contain any muscles—so that theory was debunked. Others hypothesized that the star could sense bioelectricity, similar to what is seen in sharks and even salamanders. Or perhaps the structure was some kind of supersniffer—after all, the star sits on the mole's nose. But here again, after crawling all over the star's surface with a scanning electron microscope, Catania could find no evidence of taste buds, chemical receptors, or scent receptors, which discounted any illusions about the star being some sort of supersensory radar dish, proving that it had about as much to do with sniffing as the tip of your own nose does.

(Star-nosed moles *can* smell, though, using regular ol' nostrils positioned in the middle of the star. In fact, they can even do so underwater, blowing out bubbles and sucking them back in at a frequency of about 10 times per second. And the star plays a role in this aquatic odor detection, holding the bubbles in place like a catcher's mitt. Now, you may be wondering, doesn't dirt get stuck in the mole's nostrils? Yes, it probably does, but a quick snort of air can clear a blocked nose in a way that you can't really do with an earhole.)

One thing the microscope did reveal? About a bajillion tiny domes, known by scientists as Eimer's organs. And below the skin, each of these was found to be wired with between five and 10 myelinated nerve fibers, "which are these large, fast-conducting, really important sensory inputs about the things that the mole is touching," says Catania.

Incredibly, when he sat down to quantify the star-nosed mole's sensory potential, Catania discovered that the star is actually a supernova. After days of squinting at magnified images, he had discovered about 25,000 Eimer's organs collectively hot-wired to around 112,000 nerve fibers. This information superhighway leads directly to the mole's brain.

Lest you miss how absurd such numbers are, consider that a human hand—from palm to fingertip—contains just 17,000 touch-sensitive nerve fibers. And a mole's got 6.5 times as many fibers packed into a pink sea star the size of a dime. This incredible sensory magnitude means these small, unassuming animals must possess touch perception of such high resolution that it is basically beyond our ability to even imagine.

Now, as to what all this is for, Catania says that because star-nosed moles live in moister habitats than their relatives, they have access to tons of tiny, soft-bodied creepy-crawlies that don't show up in drier dirt. Think more along the lines of pinhead-size things like springtails and mites, as opposed to comparatively gigantic earthworms. "And so one of the things that you need to do if you're going to eat small things is, you have to eat a lot of them," says Catania, "and you have to eat them quickly."

If you or I go to bed with an empty stomach, we may be grumpy about it, but we can stave off our hunger with slumber. A star-nosed mole, meanwhile, burns so much energy each day that if it doesn't eat, it dies.

And so what the mole must do is rifle through the world, rapidly touching everything it can get its probes on to see if it tastes good. Wilder still, Catania found that there is a method to this madness.

The instant a piece of the star stumbles onto a potentially yummy morsel, it immediately shifts the item down to what researchers have come to call nasal ray #11, which is positioned just in front of the mouth. This is a unique probe in the star, different from nasal ray #10 or nasal ray #12, or any of the other nasal rays, for that matter. When it comes to food, it's #11 or bust.

This particular probe, says Catania, is bristling with even more touch sensors than the rest of the star, which makes it sort of like the focus point in human vision. If #11 likes what it feels, then the food disappears down the hatch of the mole's mouth faster than you can blink an eye.

I mean that 100 percent literally. As it turns out, Catania's own high-speed videography has revealed that star-nosed moles detect, identify, and devour food in as little as 120 milliseconds—which is both faster than the human eye can blink and more rapidly than any other mammal food-handling speed ever recorded. As proof, in 2005, the Guinness Book of World Records sent Catania a framed certificate officially crowning the star-nosed mole to be the fastest eater among all mammals.

As if all this wasn't mind-blowing enough, if you were to dissect, flatten, and stain the part of the star-nosed mole's brain known as the neocortex, you could actually see that it is not-so-subtly divided into distinct sections, each of which corresponds to one of the nasal rays. In other words, when you look at one of these creatures, the wonderful weirdness you're seeing on its face is also reflected in the very structure of its brain.

"It's one of the greatest wonders of nature," says Catania. "It tells us what evolution can produce at its extremes."

CHEMICAL WARFARE IN THE DERRIERE

STRIPED SKUNK

SPECIES: *Mephitis mephitis* **STATUS:** Least Concern **RANGE:** All of North America except Alaska and northern Canadian territories **SIZE:** Up to 32 inches long; weighs up to 13 pounds **LIFESPAN:** 3 years

---

WITH CONSPICUOUS BLACK-AND-WHITE FUR PATTERNS, skunks are some of the most recognizable animals on Earth. In fact, this is by design. Because though these tiny critters usually weigh no more than a cabbage, they can hold their own against even the most ferocious of predators. And it actually benefits the skunk family to be both easily identified and long remembered. Here's how a skunk's reputation works.

Lots of mammals, including your pet cats and dogs, have scent glands in their backsides that produce stinky secretions. In those animals, scents can mark territory and allow communication within their own species. But in the skunks, these very same glands have been honed by evolution to do harm.

The precise chemical composition of the secretions varies by species of skunk, but in general, within each of the skunk's two scent glands resides an oily, yellow sulfur-alcohol concoction known as butyl mercaptan. With its two glands combined, a single striped skunk can stash up to 30 milliliters' (or six teaspoons') worth of stink.

Now, to make maximum use of their stink juice, skunks have evolved an elaborate delivery mechanism that consists of two papillae, or nozzles, that each connect to one gland. Best of all, both nozzles can move independently of the other, meaning the skunk can actually fire in two directions at once and at distances approaching 15 feet.

"They're like miniguns," says Adam Ferguson, collection manager of mammals at the Field Museum of Natural History in Chicago.

Ferguson has more experience with skunk butt nozzles than most people would ever want. Not only has he collected road-killed skunks for use as museum specimens and trapped live animals in the wild, but he has also been sprayed by skunks multiple times over the course of his career.

During most of the skunk's waking hours—which tend to be at night—its weaponized nozzles remain concealed within its butt. But when the skunk is threatened, it aims its rear end at the source of danger while bearing down on all of that backdoor machinery to the point where the nozzles actually start to peek out of the animal's body. Put frankly, if you're ever in a situation where you can see a skunk's papillae, then you've played your cards very, very poorly.

Unlike what you've seen in the cartoons, skunk spray is not a cloud of noxious fumes. "It's pure liquid," says Ferguson. "It's almost always leaving the skunk as a spray-mist." However, sometimes a skunk will forgo the mist and deliver its payload all at once in the form of a concentrated blast, like a garden hose.

"When Adam got sprayed by the hog-nosed skunk in Mexico, it was a stream right to the face," says Molly McDonough, a professor of biology at Chicago State University.

McDonough remembers canary-yellow globules of skunk slime covering Ferguson's eyeballs, so much of it that she had to scoop the goop off him and douse his face in water while he writhed in agony. After all, they were in the Mexican desert in the middle of the night—help was not on the way. And they still had to hike down a steep hillside to get back to their truck.

Ferguson and McDonough are married, by the way—a veritable skunk-research power couple. And I feel like that fact is required knowledge before you hear what happened next.

With Ferguson temporarily blinded and the hog-nosed skunk skulking away, McDonough turned and asked, "Do I help you or do I get the skunk?

Because I know you'll be fine, but if that skunk gets away, you're going to be even more upset."

Ever the professional, McDonough got the skunk.

Some Skunks Are Just Jerks

There are several different varieties of skunk on this continent. Striped skunks are the most widespread species and also the most adaptable in their diet. They eat almost anything, including trash, which is why striped skunks often do quite well in suburban and urban environments. Hooded skunks belong to the same genus as the stripeys—*Mephitis*—and are basically the Mexican counterpart to their northerly cousins. As the name suggests, in one of the most commonly occurring hooded skunk color morphs, the white coloration doesn't usually split into stripes, but covers the animal's top side like a hood.

Spotted skunks *(Spilogale)* are smaller, more nimble, and more carnivorous than the generalist striped or hooded skunks. They're also the only skunks that regularly climb trees. Spotted skunks also perform a handstand before they spray, which is just all kinds of adorable. "Spotted skunks are the acrobats of the skunk world," says Ferguson.

And then there are the hog-nosed skunks *(Conepatus),* which survive almost entirely on insects. And this makes them nearly impossible to lure into a trap for scientific purposes (on the flip side, you can catch a striped skunk easily with a can of tuna). And this is why Ferguson and McDonough were running around the desert with a hand net and a spotlight. They even have a name for the activity: skunk football.

While foxes and turkeys run at the sight of humans, skunks keep going about their business. "I think it knows it's a skunk," says Ferguson. And when you're packing chemical weapons, you simply don't have to run.

By the way, that classic black-and-white skunk coloration? It's what scientists call aposematism, or warning colors, says Ferguson. The skunk's pirate flag can even be seen in the dead of night, when skunks are most active.

Of course, skunks are not invincible, and all kinds of critters are known to hunt them from time to time, including foxes, coyotes, lynx, bobcats, mountain lions, badgers, fishers, dogs, and great horned owls. But after completing a successful kill, many predators learn that a few bites of skunk aren't worth a potentially smelly ordeal and choose to avoid the critters in the future.

That learned experience from predators might explain not only the skunk's swagger, but also how these animals sometimes win matchups they have no business winning.

For instance, in 2011, scientists documented a showdown between a full-grown, 109-pound mountain lion known as F23 and a western spotted skunk. The cat had killed a black-tailed deer three days earlier, and camera trap footage revealed that each night the cougar would return to feed on its prize. That is, until the skunk showed up and started acting like it owned the place.

As soon as the skunk arrived at the carcass, the mountain lion backed away, suggesting it may have already run into a skunk at some point in the past. Indeed, the cat gave the skunk so much deference, it watched helplessly as the tuxedo-wearing party-crasher hopped up onto the deer and started chowing down. But that wasn't enough respect for the skunk, which after a few minutes charged at the lingering mountain lion and forced it to run away!

Amazingly, this standoff lasted all night. Again and again, the cougar came back and the skunk drove it away. Even more surprising, so far as the scientists could tell, the skunk didn't even spray the mountain lion—the mere threat of stink was enough to dissuade an apex predator some 99 times the skunk's size from a hard-won food source. In fact, it's the largest size-differential "win" between any two mammals on record.

"I often say [skunks] are like people," says Ferguson. "Some of them are just jerks." Whereas one skunk allows itself to get caught without so much as a hiss, other skunks seem ready to rumble at a moment's notice. "There are definitely trigger-happy skunks," he says.

By the way, if you have had the unlikely honor of getting skunk stink all over your person—or more likely, your free-roaming pet—you can probably skip the tomato juice bath popular in conventional wisdom. For clothing and inanimate objects, Ferguson says a few weeks in the sun helps break down the chemical signatures of skunk stink. But for living things that need to smell normal sooner rather than later, you'll need the following ingredients: one fresh quart of 3 percent hydrogen peroxide, one-quarter cup of baking soda (aka sodium bicarbonate), and one to two teaspoons of liquid dish soap.

If that sounds like the beginnings of an elementary school science fair project, well, that's because it is. This combination of ingredients is scientifically proven to take the stink-tastic thiols that make up skunk musk and transform them into inoffensive acids. You're welcome!

As for the skunk couple, sometimes there's no avoiding bringing your work home with you. "There was one point where I was like, 'I think we need a separate skunk-clothes washing machine, if we're going to carry on like this,'" says McDonough.

They've also had to institute a strict no-roadkill policy in the new Subaru. But don't worry, there's always a workaround, like placing dead skunks in garbage bags, tying the bags tightly, and then cinching the end with the car's window so that the animal remains outside while you drive.

Note to self: Never let skunk scientists borrow your ride.

THE MAGICAL MARSUPIAL

VIRGINIA OPOSSUM

SPECIES: *Didelphis virginiana* **STATUS:** Least Concern **RANGE:** Southern Canada through Mexico and Central America **SIZE:** Up to 16 inches long; weighs up to 6 pounds **LIFESPAN:** ~2 years

---

AT FIRST GLANCE, opossums seem normal enough.

They're furballs, which places them squarely among the mammals, of which North America has many. They have pointy, rodentlike snouts and a gray-and-white color pattern that looks like it would be highly sought after if you were breeding purse dogs. Virginia opossums are common enough pretty much everywhere in North America except the Rocky Mountains, eat just about anything they can get their fingers on (including poisonous toads and venomous snakes—opossums are resistant to copperhead venom, among others), and generally just aren't that difficult to spot. This despite the fact that they tend to be nocturnal.

"Most people encounter them in their trash cans or dead on the road," says Robert Voss, opossum expert and curator of mammals at the American Museum of Natural History in New York City.

But the closer you look at an opossum, the more things start to diverge from all the animals you're used to. Foxes and raccoons have luxuriously fluffy tails, but the opossum has a naked, scaly rear-end appendage more reminiscent of a muskrat's. Bobcats and groundhogs look like they wear fur-lined gloves on their paws, but opossum fingers and toes are more akin to bubble-gum-pink mini-sausages.

Of course, the one thing most everyone knows about the Virginia opossum is that if it gets into a scrap, it can keel over and play dead. Scientists call this death-feigning, or thanatosis, and while there's evidence that some of the other 125-ish species of New World marsupials found in the Western Hemisphere can do it, too, so far, it seems like the Virginia opossum is the most awesome possum-player of them all.

In his book, *Opossums: An Adaptive Radiation of New World Marsupials,* Voss describes death-feigning as "a highly stereotyped behavior that is consistently exhibited only when the animal is seized by the neck and violently shaken, typically by a dog." This means that the opossum is not like a fainting goat. It does not simply fall over every time someone shouts *BOO!* but typically reserves its secret weapon for only those last-ditch, all-or-nothing situations that are likely to end in death.

When opossums are backed into a corner, something in their brain triggers a cascade of biological responses. First the animal falls onto its side and its legs and tail go stiff. The corners of its mouth tighten and drool starts gushing out. The eyes remain open, but the rest of the creature appears to go positively catatonic.

Weirdly, Voss notes that beneath the act, the opossum is not in some sort of protective fugue state but instead is fully conscious. And studies have confirmed this by showing that opossum heart rates and brain functions continue as normal even when in the throes of thanatosis.

Strangest of all? We're not totally sure why any of this works. After all, when a predator latches onto a small mammal with the intent to kill, eventually that animal stops moving—which is exactly what playing possum achieves. The difference may simply be that when an opossum gives up without a fight, so to speak, it confuses the predator and short-circuits its instinct to attack. Though it's probably also no coincidence that opossums poop themselves and emit a stinky green goo out of their anal glands as part of their death dance. Whether there's something in those secretions that deters a predator or makes it lose its appetite, we know not.

The gambit doesn't always work, of course, and sometimes a hungry coyote, mountain lion, or neighborhood dog will muscle through the stench and kill the opossum anyway. But when it does succeed, you can't help but admire the gumption. The opossum knows it can neither fend off the predator nor escape by some other means. So what does it do? The opossum renders itself so thoroughly inedible that the predator just gives up an otherwise free meal.

Bravo, indeed.

Of Placentas and Pouches

Predator-defusing superpowers aside, there's another big, irreconcilable difference that sets Virginia opossums apart from literally every other mammal in Canada and the United States. And it all comes down to a disc of extremely important and surprisingly mysterious tissue known as the placenta.

For humans, beavers, elephants, tigers, narwhals, and the vast majority of mammals on Earth, the placenta serves as a pregnancy-induced portal between the mama mammal and the offspring growing inside her uterus. This is why scientists call us, as well as all those other critters, placental mammals.

On the other side of the equation are the marsupial mammals, which include opossums, as well as many species found in the South Pacific, including koalas, kangaroos, wombats, and Tasmanian devils. (There is also a third group of mammals that resides outside both categories. They are known as the monotremes, and they include only a few species of echidnas, as well as the duck-billed platypus. These animals, rather than bear live young, lay eggs and thus are mammalian outliers in their own wacky right.)

Interestingly, marsupial mammals do not lack a placenta entirely, says Voss. It's just that it lasts only a few days, rather than the months or years seen in placental mammals. And the reason for that is that marsupials have supercharged pregnancies.

You probably know that humans give birth after about 40 weeks of gestation. African bush elephants take an even longer approach to pregnancy, staying pregnant for around a hundred weeks, nearly two full years. House cats have their kittens in a little over nine weeks. But the Virginia opossum, which can be roughly the same size as a large house cat, gives birth after just 12 *days*. This is one of the shortest pregnancies of any mammal.

But here's the thing: The babies, or joeys, that come out aren't even close to being done developing. They are bald, blind, and deaf. Their skin is so thin as to be nearly transparent. Their skeletons don't even contain bone yet and are composed entirely of cartilage. Honestly, these things barely have a central nervous system or functioning digestive organs. Even their brains are only 9 percent complete.

The one thing the joeys do have is claws. Huge, gnarly claws. Claws that they use to haul themselves out of their mother's womb and into her pouch, which is a flap of skin located on the lower abdomen.

Now, that might not seem like that big of a deal, since the distance between pouch and birth canal is just a few inches—barely more than a smidge—but consider that each joey is just the size of a grain of rice or a small bean at birth, says Voss. Scaled up, the journey would be like if human babies, in the seconds after they were born, had to crawl into a crib six feet away. Oh, and did I mention the crib would be on the ceiling? Because the mothers give birth while sitting on their haunches, which means the joeys have to *climb* to get where they're going. With knives for fingers, the joeys grapple and paw their way through Mom's fur, performing a sort of overhand stroke.

Once inside the pouch, which the mother can open and close with a sphincter, the joeys must find a nipple and hang on like their lives depend on it. This is not a figure of speech. The mother opossum has just 13 nipples, all of them arranged in a U shape, with one in the middle. She might give birth to dozens of babies, but she can only nurture a maximum of 13 young, so the infants are sometimes competing with one another for survival.

And not all the nipples work all of the time, notes Voss, so finding the right mammary gland is also sort of like Russian roulette. Claiming a spot is so crucial, in fact, that when the joeys latch on, their lips actually fuse closed around the nipple. If you try to remove a joey in this early state, you run the risk of tearing its lips.

Over time, the mama opossum's teats will stretch and elongate up to 35 times their original length, which allows the joeys to explore the pouch without letting go of their lifelines. No wonder people once believed opossum babies sprouted directly out of the mother's nipples, like grapes on a vine.

Yes, the marsupial way of life may strike North Americans as odd, but it's no more correct or normal than that of placental mammals. After all, each reproductive strategy has worked well for around 100 million years, and modern-day opossums are as wily and cunning as any creature you'll find in the animal kingdom. But now that you know what's going on inside that pouch, I'll bet you never look at a dumpster-diving opossum the same way again.

THE SUPER-SENSORY BLOB

WEST INDIAN MANATEE

SPECIES: *Trichechus manatus* **STATUS:** Vulnerable **RANGE:** Eastern coastline of North, Central, and South America **SIZE:** Up to 13 feet 6 inches long; weighs up to 3,500 pounds **LIFESPAN:** 60 years

---

THE HEAVIEST MANATEE on record weighed more than a black rhino. No land-living creature in North America even comes close. Even at its more typical weight of around a thousand pounds, the manatee is nothing short of gigantic.

The Sirenians carry it well, though. (Yes, the manatee's order gets its name, Sirenia, because early sailors somehow mistook the animals for mythical Sirens.) Gliding through coastal waters like overfilled zeppelins, manatees are primarily herbivores, although they may vacuum up the occasional fish or crab. Which means they acquire all those pounds by slurping down pond weeds, wild celery, mangrove leaves, water hyacinth, turtle grass, shoalgrass, widgeon grass, and even something known as manatee grass.

Interestingly, lots of these plants are full of silica, which you may know as a mineral component of quartz, glass, and sand. Silica performs many useful functions inside plants, including making their cell walls more rigid, but it also makes these plants an absolute torment on the teeth of animals that try to eat them. Fortunately, manatees have engineered a clever solution to all of this dental wear and tear.

Unlike humans, who get just two sets of teeth, the manatee's mouth is an ever evolving plant-processing apparatus.

"Manatees actually have a tooth-generating organ in the back part of their jaw," says Roger Reep, a manatee expert and professor emeritus at the University of Florida.

Manatees have neither incisors nor canines—only molars. And as that tooth organ churns out new molars, it pushes ever so gently on the old ones, moving them forward at a rate of about one millimeter per month, says Reep, who is also the author of *The Florida Manatee: Biology and Conservation*. Nerves stretch, sockets slide, and eventually, the worn-down chompers up front plop out and get replaced by a new class of gnawers. Scientists call the phenomenon marching molars, but you could also think of it sort of like a conveyor belt of teeth. (Fascinatingly, elephants have marching molars, too, which makes a little more sense when you learn that they are also one of the manatee's closest living evolutionary relatives.)

Manatees descend from four-legged land walkers, but today those hind limbs have all but disappeared, leaving a huge slab of a tail in their place. Flippers on either side conceal handlike bones. The rest of their skeleton has changed over the millennia, too. Whereas our largest bones contain spongy marrow in the middle, manatee bones became more dense, heavy, and weirdly solid. Scientists believe this extra weight helps the marine mammals sink down to the ocean floor or riverbed, where their food grows.

Another manatee fact, and one you can't learn just by looking at them: If you were to crack open the skull of one of these behemoths—though, please don't—you'd find a pink wad of Silly Putty–like brain that is surprisingly tiny. We're talking the size of a grapefruit for an animal that can outweigh a Toyota RAV4 sport utility vehicle.

Weirder still, the manatee's brain has no wrinkles.

Now, this has led many—scientists included!—to believe that manatees are rather stupid. And this reputation has not been helped by the fact that the animals move slowly and aren't super great at getting out of the way of motorboats, which is why many manatees have scars all over their backs. Throw in a plant-based diet similar to that of grass-munching cattle, which, again, are not creatures known for their remarkable brainpower, and you wind up with a nickname that inspires exactly zero intellectual enthusiasm: the sea cow.

Moo. But also, *boo.* Because while manatees might seem like some of planet Earth's more, shall we say, carefree animal species, they have a few special skills that ought to earn them a more impressive reputation.

The Manatee's Secret Sensory Superpower

Let's go back to the brain for a second. Sure, the fact that manatee brains lack folds seems like a red flag. But birds have smooth brains, and yet crows, ravens, and parrots are heralded as smarty-pants species. Plenty of mammals have smooth brains, too, including bunnies, squirrel monkeys, and koalas, and I don't see anyone putting a dunce cap on them.

In fact, Reep says brain shape and size aren't necessarily good ways to measure an animal's complexity. And despite the fact that manatee brains are smaller and simpler than we'd expect, that doesn't mean that there's anything wrong with them. "They have a very nicely structured brain," he says. "It has all the right layers and everything."

For some animals, being slow and steady isn't an option, says Reep. The cheetah, falcon, and marlin would starve if not for speed. On top of that, many animals struggle against food scarcity, which means they must work hard to find and subdue their lunch. But a manatee's meal is relatively abundant and can't run away. That means all they need to do is show up and chew.

But there's another factor that pushes brain development—the threat of being eaten. A deer at the edge of a cornfield constantly scans for impending doom, because every moment spent nibbling exposes it to the risk of becoming food for something else. And that kind of fear, it changes you—makes you wary, makes you wired.

"But there's nobody trying to eat a manatee. They don't have natural predators," says Reep. "There are occasional shark or alligator bites, but they're accidental, as far as we know."

With no reason to hurry, manatees were able to opt out of lots of costly brain machinery other animals need to survive. But that doesn't mean manatees lack evolutionary bells and whistles.

Look closely at a manatee's mug and you'll notice it's covered in stout, stiff hairs, rather like those of a walrus. These bristles actually help manatees manipulate food and move it into their mouths. But each hair is also rigged up with extremely sensitive nerves, which means the manatee's mustache isn't just for looks. It's a *sensory organ*.

"If they're investigating a novel object, like the leg of a swimmer or a crab trap, they will tactically inspect it by direct contact with the sensory hairs," Reep says. What's more, while bristles are concentrated on the manatee's face, the animals also have around 3,000 hairs scattered across the rest of their body. And each one is hooked into the central nervous system with a sensitivity twice that available to people who can use their fingertips to read Braille.

To learn more about these hairs, Reep and his colleagues trained manatees to touch their noses to a target when they detected a small movement in the water. Then, they shaved off the manatee's bristles with a beard trimmer and ran the experiment again. Without body hair, the manatees' ability to sense vibrations was clearly dampened—but it didn't disappear completely, perhaps because human beard trimmers couldn't quite shave the marine mammals clean. But that in itself is intriguing. Maybe a bit of supersensory stubble is all manatees need!

It's important to note how uncommon it is for a mammal to have sensory hairs, which are technically called vibrissae, all over its body. Most just have a few here or there. Squirrels have vibrissae on their elbows, for instance, probably to help them navigate the forest canopy. Cat facial whiskers supplement short-distance vision, whether it's to help squeeze through a cat door or the low brush found in a forest.

"We don't really have sensory hairs, as humans," says Reep. "But manatees? This is the only kind of hair they have."

Armed with the equivalent of thousands of tiny radar dishes, manatees can feel slight changes in water flow from any direction. This likely helps the animals detect changes in tides and water currents, aiding with navigation. Reep says it's also possible the manatees can feel the vibrations that bounce back from their own movements in the water, which would mean the animals have evolved a type of crude sonar—though this has not yet been tested.

As it turns out, manatees are only as smart as they need to be. And it's safe to say the same rule applies to scores of other animals we tend to think of as lazy or dull (like sloths, for instance). When we get out of our human-centric worldview, we can appreciate manatees for what they are—gentle giants that have persisted for millennia while avoiding conflict with their neighbors and eating food most other animals don't want. And if that weren't enough, manatees can tell the turning of the tide using only their back hair.

Oh, and the fact that they can't get out of the way of speedboats? Manatees have been around for *50 million years*. The first speedboat was invented in 1888. That means speedboats have been present for just 0.0000027 percent of the manatee's existence. So maybe give them a minute before expecting them to adapt.

THE SURVIVOR

WHITE-TAILED DEER

SPECIES: *Odocoileus virginianus* **STATUS:** Least Concern **RANGE:** Alaska to Peru **SIZE:** Up to 3 feet tall at the shoulder; weighs up to 300 pounds **LIFESPAN:** Up to around 15 years, but usually between 1 and 4 years

---

BENEATH A QUIET FACADE, deer boast panoramic vision, superpowered scent detection, and a host of behaviors honed over the millennia to ensure that they do not become someone else's supper.

Today, we might look at deer and see innocence. But closer inspection reveals an organism shaped by saber-toothed cats and packs of dire wolves. A grizzled survivor fueled by little more than twigs and grit. A species that managed to colonize every habitat from British Columbia to Peru's border with Chile.

"We kind of take them for granted as some sort of victims in the animal world, because they're a prey species," says Jeannine Fleegle, a deer biologist for the Pennsylvania Game Commission.

As one of the scientists tasked with managing the Keystone State's deer herd, Fleegle says she's seen deer running through the forest with their guts hanging out—usually the result of a hunter's misplaced shot. It's also not uncommon to encounter these animals missing entire limbs. "There are oodles of three-legged deer running around out there," says Fleegle. "One of those legs is expendable."

Each winter, deer routinely endure fasts lasting up to a month at a time when food becomes scarce. In spring, deer sometimes gobble up baby birds out of low-hanging nests and snakes right off the ground for a little boost of nutrition. And a doe won't think twice about stomping your golden retriever to death with her hooves if the pooch gets too close to her fawns.

Bucks, with their giant, majestic antlers, receive the most attention, perhaps because their head-banging bouts of dominance are a sight to behold. But what many people don't realize is that these battles can take a dark turn.

Every so often, two bucks' antlers become locked together. When this happens, it's common for both animals to die, either from starvation, predators, or injuries sustained in the fight. But sometimes, one of the males survives and goes on trotting around the woods with the rotting skull of his foe securely fastened to his forehead. This has been documented more than once, by the way. Sometimes, three deer get locked together. There's also a record of one buck who somehow survived after its antler-locked adversary was killed and eaten by coyotes.

In other words, Bambi this is not.

Of the three species of deer in North America, the white-tail is the widest ranging, inhabiting 18 countries and a breathtaking 78 degrees of latitude. In fact, the Arctic Circle seems to be the only thing that can stop them. White-tailed deer share habitat with anacondas and jaguars in the Amazon rainforest, grizzlies in the Rockies, Mexican wolves in the American Southwest, and coyotes in suburban backyards across the continent. "I mean, the only place that deer cannot live is in water," says Fleegle.

The Wonders of Antlerogenesis

Let's begin with a simple clarification: Rams, antelopes, and rhinos have *horns*. Deer have *antlers*. This might seem like pedantry, but the difference is more important than you might think.

Horns are permanent structures built out of the same stuff as your hair and fingernails, and just like those structures, they grow slowly and continuously over time. Horns are sort of simple, biologically speaking. And they make logical sense. But antlers? Antlers are just plain *weird*.

Unlike horns, antlers are temporary, rapidly growing organs. Each year, a male deer grows a new set, and then, after a while, he drops them like a fading fad. This would be like if you spent a year growing a new pair of kidneys out of your face, only to rip the things off and build another set from scratch the next year. Oh, and those kidneys are full of bone.

Here's how it works. On their foreheads, male deer have a pair of raised knobs called pedicles. Each pedicle is covered in special stem cells that allow for rapid tissue growth, usually in response to waning levels of daylight. Through a bit of biological magic scientists still don't fully understand, living flesh sprouts from this pedicle at a rate of up to 1.5 inches each week. This makes antler growth one of the fastest rates of organogenesis found in the animal kingdom.

When they first sprout, antlers are formed out of cartilage, with a thin, fuzzy layer of skin coating the outside. This is known as velvet, and it's actually a membrane that protects the coursing veins beneath. Remember, at this stage, antlers are alive, warm to the touch, and stuffed full of blood, which carries protein, phosphorus, and calcium that help convert springy, sensitive cartilage into battle-worthy bone.

Because a buck's daily diet can't quite cover such rapid growth, his body will actually leach minerals out of non-weight-bearing bones, such as the ribs, to support the growth of the new organs. Scientists call the process cyclical reversible osteoporosis.

Eventually, the deer's body cuts off the gravy train of blood and nutrients, the velvet dies, and the whole structure starts to calcify and harden. It also gets really itchy, and bucks can be seen rubbing their velvet off on tree trunks and branches—much to the fury of gardeners and groundskeepers everywhere.

Once scoured clean, antlers become sensationless weapons. They can clack, crash, and gore. They are anchored to the deer's skull and capable of withstanding massive amounts of blunt force trauma.

But antlers are also a liability. In addition to potentially getting stuck on another deer, antlers can also become entangled in briars or branches overhead, as well as wire fencing, Christmas lights, and garbage bags. All of which is to say, there is an advantage to getting rid of your head ornaments once the breeding season is over.

Remember how it's called cyclical reversible osteoporosis? After mating season, a male deer's body will actually begin to suck some of those nutrients back down into its body. This weakens the antlers to the point that, one day, they just pop off—usually within hours of each other. Indeed, the separation of tissue between the antler and its pedicle has been described as the most rapid deterioration of living tissue known by scientists.

Now, in the '60s, '70s, and '80s, scientists conducted a series of experiments to test the limits of antler growth in roe and fallow deer, which are species native to Europe. After sedating the animals, the scientists surgically removed portions of their pedicles and then inserted them as a graft beneath the skin of the lower legs.

After just one month, as deer in the control group started to form budding antlers on their heads, the deer with pedicle grafts in their shins also began to show small bumps of growth. Over time, the fur covering the swellings started to thin out and look identical to antler velvet. By late summer and early fall, the growths even shed their velvet to reveal small, honest-to-Betsy antlers. On legs. Leglers!

And get this—just like real antlers, the leglers popped off after the breeding season ended. And then, just like real antlers, they started growing again the next summer. Some of the leglers were even bigger in their second season, with one protuberance reaching nearly three inches in length!

Scientists have also transplanted pedicle tissue and skin from a red deer onto the foreheads of lab mice. Unbelievably, these transplants grew, too, resulting in a bunch of rodents with giant, round protuberances of bone and cartilage growing out of their faces. Kind of like if you asked a plastic surgeon to slip an eggplant under the skin of your brow.

"Regardless of where they are, those cells are going to grow an antler," says Fleegle. "It's what they're programmed to do."

You can chop off a deer's antler growth site, mince it up, and put it back, and an antler will still form. A female deer can even grow antlers when injected with testosterone, a hormone produced by the testicles in males. Injuries to a pedicle result in misshapen antlers, and those injuries are "remembered" through time, yielding irregular antlers year after year.

If a male deer is castrated as a fawn (either by scientific manipulation or natural injury), then it will never grow a set of antlers. But if the testicles are damaged or removed after a buck has already begun to grow a set of adornments, the deer will remain in velvet for the rest of its life. The skin never dies, nor do the antlers cast off after the mating season is over. They just keep growing. New tines pop up around the old ones each year, until the deer's rack starts to look like a mound of coral. Or cacti, as deer with this unfortunate condition are known as cactus bucks.

Some think the key to regenerating human organs and limbs may be locked inside antler cells. One day, antler research could even help people with damaged spines regrow nerves and walk again. Which kind of makes you wonder, what other potential medical or scientific breakthroughs might be bounding around our backyards?

THE GLUTTON

WOLVERINE

SPECIES: *Gulo gulo* **STATUS:** Least Concern **RANGE:** Northern regions worldwide, including Canada, Alaska, and parts of the American Northwest **SIZE:** Up to 3 feet long; weighs up to 40 pounds **LIFESPAN:** Up to 13 years

---

FRIENDS, I WOULD LIKE TO INTRODUCE YOU to the wolverine, whose scientific name, *Gulo gulo,* translates to "the glutton." Or rather, *Glutton glutton*—the glutton so nice, they named it twice. These hell-raisers look like flattened bears but are actually the largest terrestrial member of the Mustelidae, or weasel, family. Kings and queens of the weasels, if you will.

In some circles, wolverines are known as skunk bears, due in part to the way they arch their hindquarters and also because of the sinus-singeing, yellow-brown discharge they sometimes emit from their backsides when spooked.

Wolverines can also produce some seriously scary vocalizations. For instance, if you ever happen upon two siblings in the midst of a playful tussle, you might think they were actively trying to tear each other's heads off. And, of course, a hefty dose of gruffness and ferocious individuality is no doubt why a certain beloved comic book antihero shares his name with these sassy weasels.

"So yeah, wolverines have a big reputation," says Mirjam Barrueto, a wildlife biologist at the University of Calgary. A reputation they can back up, by the way.

Wolverines have been seen chasing black bears into trees and even chasing brown bears off a kill. One time, a Native American hunter even alleged that he'd seen a wolverine latch onto the throat of a full-grown polar bear—the largest land carnivore on the entire planet—and hang on until that ice bear was no more.

Despite all the street cred, Barrueto says the chances of one of these chocolate-and-peanut-butter-colored mammals hurting a human are pretty low. "Mostly, it's bluff," she says.

In any event, before you go to bed dreaming of demon weasels, it's good to remember that even the largest male wolverine on record weighed in at just 55 pounds. And most males top out at around 40 pounds, with females only around 26 pounds. "That's kind of like a midsize dog," says Barrueto. "They're not really an animal that could do much damage to a person. Unless you lie on the ground in front of a wolverine and don't move." Maybe then a wolverine would try to eat you, she says.

Wolverines stand no taller than a footstool yet manage to eke out a living in one of the most unforgiving environments on the planet. They don't hibernate, they don't make friends, they give birth in the dead of winter, and they live within a whisper's distance of freezing or starving to death pretty much every day of their lives. And all of these strategies have served the gluttons well over the past several million years.

But then, along came humanity.

Save the Female Wolverines, Save the World

Because wolverines inhabit remote habitats, such as boreal forest, tundra, and alpine areas, you might think they'd be more protected from people than most animals. But these sorts of places have very few tasty treats per square mile, which means wolverines must cover a lot of ground in search of sustenance.

Like seek-and-destroy Roombas roaming through the mountains, wolverines eat whatever they can find, be it dead or alive. Each wolverine maintains a home range that is furiously defended from interlopers and can cover vast, unthinkable distances. One female in Montana was estimated to lord over more than 370 square miles, or an area more massive than the entire country of Grenada. Another male did his best impression of one of Jack Kerouac's beat poets, traipsing 250 miles across the Grand Tetons in just 19 days.

"Some wolverines never establish a proper territory," says Barrueto. "They just float for years, on the margins and intruding into other wolverines' territories."

Of course, you and I wouldn't do well if someone dropped us on the side of a mountain and told us to take a hike, but wolverines are built for life on the lam.

Wolverine paws are extra large and act like fur-lined snowshoes, which allow the gluttons to plod along the backcountry in conditions other forestgoers find impassable. Indeed, while it's true that wolverines might attack a full-grown caribou or moose, they are more successful when snowdrifts make it difficult for those animals to escape. Each of those paws also has long claws, which give wolverines the power to dig through snow and soil until they reach sweet, defenseless hibernators, such as marmots, just trying to snooze through the winter.

In their mouths, wolverines have a rather strange arrangement in which the last molars on the roof of their mouths are turned 90 degrees—a condition scientists believe helps the wandering weasels crunch through bones and extract all the nutrients they can from a corpse.

The good news for scientists is that, because food is often so hard to come by in their natural environs, wolverines are actually quite easy to lure into camera trap stations.

To entice wolverines out of the woodwork, Barrueto flies into the backcountry by helicopter, then skis and snowshoes her way into the wild. Then she constructs a sort of tree house out of 2 × 4s and circles it with barbed wire, alligator clips, and motion-activated cameras. Finally, she uses a steel cable to hang a nice, fat, frozen and skinned beaver carcass just out of reach of the contraption.

Grizzlies, Canada lynx, pine martens, and even flying squirrels will come from far and wide to nibble on the beaver, but so will wolverines. And the more times the wolverines attempt to carve off a mouthful of frozen flesh, the more data Barrueto collects. She gets DNA from hair caught in the barbed wire and clips, and she can spot unique patterns on the animals' chests to identify individuals by sight. Sometimes, the cameras even catch a glimpse of a female wolverine with elongated nipples—a sure sign that a kit, or baby wolverine, is nearby.

While this may sound like an invasion of privacy—scoot through this barbed wire and show us your nipples to get a treat!—the sampling stations are actually a handy, noninvasive way to sample reclusive wildlife without stressing them out, says Barrueto.

With all this data, Barrueto showed that while protected areas held three times as many wolverines as unprotected areas, all it took was just a little bit of human disturbance to make the wolverines there go into hiding. For instance, just three groups of humans passing through an area in a period of two weeks caused wolverine sightings on the cameras to dip. At the same time, wolverine populations within those same protected areas actually decreased by 39 percent over the course of nearly a decade—likely as a result of legal trapping combined with the effects of human disturbance.

Worst of all, female wolverines appear to be much more bothered by human presence than males. And that's a big problem. "Without females, there are no babies, right? Without babies, there's no population," says Barrueto.

Discounting human trappers and the occasional collision with a vehicle, not much out there dances with a fully grown wolverine. Wolves and cougars nab a wolverine every once in a while, but by and large, once a wolverine becomes an adult, it lives until it dies of starvation or natural causes. But the kits (which, by the way, are born blond) are highly susceptible to predators—including other wolverines.

All of this adds up to a rather slow reproductive rate for the species. And under normal circumstances, that would be okay, because wolverines are solitary animals, and the females especially live by *Highlander* code—there can be only one (in a given area).

But when you add in threats like logging, loss of prey base, human recreation, encroaching developments, and trapping—factors that led Canada to list the wolverine as a species of "special concern" in 2018 and the U.S. to follow suit with a "threatened" status in 2023—this extremely delicate balance starts to tip into the red. And that's what leads Barrueto to believe the future of the wolverine hinges on our ability to make room for lady skunk bears. To help, she's created the Female Wolverine Project—a scientific initiative geared toward the better understanding of these animals and how to keep them on the landscape.

"The females are really like this beautiful little flower," she says. "And we need to protect it."

By the way, Barrueto says that protecting female wolverines isn't just good for skunk bears, but also for grizzlies, caribou, and many other specialist species that share their world.

Wolverines have earned their tenacious reputation for flourishing in the bleak midwinter, on the fringe, and in the middle of nowhere, in places that supply clean air, cold water, and a respite from the developing world springing up all around us. But they are also fragile. And interestingly, this is where Barrueto sees hope. For as sensitive as wolverines are, and as colossal as some of the risks against them can be, these landscapes still exist, she says.

And so long as they do, there will be room for wolverines to wander.

Northern spotted owl
(Strix occidentalis caurina)

PART II

BREATHTAKING BIRDS

THE BACKYARD BRAINIAC

AMERICAN CROW

SPECIES: *Corvus brachyrhynchos* **STATUS:** Least Concern **RANGE:** Continental U.S. through central Canada **SIZE:** Up to 20 inches long, 3-foot wingspan; weighs up to 22 ounces **LIFESPAN:** Up to 17 years

---

CROWS ARE EASILY IDENTIFIED by their jet-black feathers, bills, eyeballs, and feet—and by the way they are probably watching you as we speak, studying you, trying to determine if you're a friend or foe.

Lots of animals are wary of humans. But crows? They never forget a face.

In an experiment that involved trapping, banding, and releasing American crows while wearing a caveman mask, researchers learned that the birds remembered the "face" of the trapper. How do we know? Because anytime the scientists busted out the caveman mask and went for a stroll in the area where the birds had originally been trapped, all the crows around started screaming their heads off.

Scientists call this behavior scolding, and crows do it to people or animals they consider to be a threat. To be clear, the crows did not scold just any ol' researcher wearing a mask. For instance, the crows were also exposed to a person wearing a mask of former U.S. Vice President Dick Cheney, but because that mask did not capture any crows, the animals paid it no mind. Nor do crows regularly scold nonmasked humans.

Add it all up, and the scientists believe the animals were reacting to a very specific danger: the face of a known crow-napper. But wait. It gets better!

In a follow-up study, researchers discovered that knowledge of the caveman mask had spread throughout the local crow population. In other words, crow mothers and fathers taught their crow chicks to recognize and fear the dreaded caveman. Scientists call this sort of knowledge transfer vertical social learning. Similarly, crows that had never been trapped themselves learned to fear the caveman by watching other crows go through the ordeal, which is horizontal social learning.

Best of all, when the scientists went back to the area five years later, the crows still remembered the Legend of the Crow-napping Caveman.

Build Your Own Crow Army

Kaeli Swift is probably one of the most well-known crow evangelists on Earth, and even though she studies crow thanotology—or how crows interpret or react to death—she thinks these deeply curious birds offer wildlife watchers a sort of total package.

"We're so lucky," says Swift, who is an avian behavioral ecologist at the University of Washington. "Of all the animals to just have in most people's backyards, it's *this one*."

Think about it, she says. The world is becoming increasingly urbanized, and people are feeling more and more disconnected from wildlife and nature. At the same time, most of us will never get to see the great wildebeest migration across Africa's Serengeti or the way monarch butterflies huddle in the Mexican highlands by the hundreds of thousands.

But crows are pretty much everywhere. The American crow inhabits virtually every square mile of the continental United States, from backwoods to back patios, as well as parts of northern Mexico and southern Canada. And that's just one crow species among many! Fish crows live in the American southeast, Tamaulipas crows inhabit the coast of the Gulf of Mexico, and even far-flung Hawaii has its own crow species, which is nearly extinct, called the ʻalalā. Elsewhere on Earth are Jamaican crows, eastern jungle crows, New Caledonian crows, violet crows, Somali crows, and carrion crows, to name but a few.

Suffice it to say, crows are a varied and widespread branch of bird-dom, in part because these animals are as adaptable as they are clever. Unlike most critters, crows don't just survive in the presence of humans, they thrive. In fact, some of the best places to see crows are now deep in the hearts of human cities, thanks to a behavior known as communal roosting.

Unlike nesting, which is when birds build structures to incubate eggs and raise chicks, communal roosting is sort of like a giant sleepover. Only instead of half a dozen kids staying up late and making a ruckus, we're talking about tens to hundreds of thousands of loud-mouthed birds.

In some places, crows sleep in urban communal roosts year round. In others, their activity peaks over winter, perhaps because cities and all their concrete trap and radiate heat (aka the urban heat island effect) and that helps keep crows snug through the coldest months.

As you might imagine, not everyone is a fan of communal crow roosts, mostly because the birds poop in great quantities, pose a risk to nearby airports, and generally display a lack of respect for quiet hours. But Swift doesn't really see it that way.

"It's just an incredible experience," says Swift. "While some people look at it as being sort of a nuisance, really what it is is an opportunity for city-dwelling people to see an animal phenomenon on a potentially enormous scale."

Also, all those quorks and caws that drive neighbors nuts? They're highly advanced communications that can vary by individual and that, frankly, we still don't fully understand! In fact, scientists are recording new crow sounds all the time. "So if you ever hear what sounds like water drops or the rattle of a creaky door, any of those things could be the crows talking," says Swift.

Apart from being easy to find and prone to grand assemblages, the third thing that makes crows an underappreciated creature is the way these birds often do things that are eerily humanlike. "They pair for life, maintain territories most years, and they often have extended family groups," says Swift. "That describes most people."

Most bird parents kick the kids out of the nest once they're old enough to fly, and, indeed, some parents simply fly away and never come back, leaving the next generation to figure life out on its own. But crow offspring often stay tight with their families, sticking close to their parents for as long as five years after birth.

The kids earn their keep, too! Once they can fly, every crow in the family group contributes by gathering sticks and nesting materials each spring, bringing Mom food while she's incubating the eggs, standing guard over the nest, and feeding their siblings once they hatch. Sometimes, juvenile crows will even leave the family and stop back in for a visit years later. Almost like they're home visiting from college.

Of course, New Caledonian crows—which are closely related to American crows—are famous for being able to assemble compound tools in order to solve puzzles. And all crows have weirdly big brains relative to their body size, which is very similar to our own anatomy, says Swift. No wonder some scientists liken the species to flying monkeys.

Crows are also one of the few animals thought to engage in play. Whether it's knocking each other off a wire, surfing on updrafts, riding a piece of plastic down a pitched roof, or simply rolling around in the snow, such behavior joins crows with only around 25 other bird species (out of the 10,500 known) that will, from time to time, get a little silly.

Despite all of the above, plenty of people still associate crows with bad omens, a lot of which stems from a time of great violence in the world—namely, during the Crusades, when Europe and the Middle East were inundated with dead bodies from war and pestilence. "And those dead people were getting eaten by crows," explains Swift.

Of course, crows aren't inherently any more good or evil than bunnies, buffalo, or bears, but unlike most animals, crows do seem to have a special understanding of death.

When crows spot a crow that's been killed—either in traffic or by a predator—the birds will gather around it in a phenomenon some have likened to a funeral. And while Swift can't speak to whether they are mourning, she has identified one thing they certainly are doing: They're *learning,* she says.

During her doctorate program, Swift tested this question by again exposing crows to people wearing masks. This time, however, the people were also sometimes holding a dead, taxidermied crow or pigeon. When researchers were holding stuffed crows, the observant crows quickly learned to associate the masked humans with danger.

What's more, another study used brain scans to reveal that when a crow sees a human holding a dead crow, the hippocampus—the region of the crow's brain that is associated with learning and memory—lights up like a Christmas tree.

"People find dead crows and pick them up and then they email me and say, 'Why won't the crows leave me alone?'" says Swift. "And I say, 'Well ...'"

Rather than making enemies of your neighborhood crows, you might just try making friends. According to Swift, you can ingratiate yourself to the local family group by offering no more than five pieces of high fat or high protein pet food once a day (more could mess with the crows' normal foraging, in addition to attracting rats, ants, and the like).

In time, the birds will remember you as their benefactor and even croak or come flying when they see you approach. "You're really getting to know individuals," says Swift. "That's why I think feeding crows is a more engaging experience than putting up a bird feeder."

Just promise me you'll use your newly begotten crow army for good.

THE PATRIOTIC PIRATE

BALD EAGLE

SPECIES: *Haliaeetus leucocephalus* **STATUS:** Least Concern **RANGE:** From Alaska to northern Mexico **SIZE:** Up to 37.8 inches long, wingspan up to 6 feet 6 inches; weighs up to 13 pounds **LIFESPAN:** Up to 38 years

---

THE BALD EAGLE is an utter beast of a bird.

With a razor-sharp banana for a beak and a wingspan surpassing six feet, there aren't many flying things on this continent that can top it. Heck, bald eagles may be even more impressive when sitting on the ground, where they approach heights of three feet or more—or about as tall as a human toddler. And if you ever get to watch one of these white-headed, brown-bodied predators pierce a still-gasping salmon with talons the size of your thumb and then shred that fish to bloody ribbons, well, you'll understand why this descendant of the dinosaurs is a raptor to be reckoned with.

By the way, those talons? When they close, they actually lock in place, thanks to a series of tendon notches that allow them to ratchet their grip tighter and tighter. All in, a bald eagle can cinch down on its prey with a clutch roughly 10 times stronger than the human hand is capable of.

Now, despite being impressive physical specimens, there is one bald eagle trait that doesn't live up to what you might have seen on television—the patented bald eagle scream. In fact, any birder will tell you that the sky-rending screech that accompanies bald eagles in most media depictions actually belongs to a red-tailed hawk. So, what do bald eagles sound like?

"I always think that bald eagles sound like they're giggling," says Janet Ng, a wildlife biologist for the Canadian Wildlife Service. "A whistling kind of giggle."

This isn't the only thing we get wrong either. For though bald eagles are capable predators in their own right—they sport telescopic vision and can see in ultraviolet—they're not above letting other animals do their work for them. In fact, Ben Franklin—scientist of the American Enlightenment and one of America's founding fathers—famously argued against adopting the bald eagle as a national symbol for this very reason. The "Bald Eagle...is a Bird of bad moral Character. He does not get his Living honestly...[he] is too lazy to fish for himself," wrote Franklin. And this is actually true.

"Bald eagles are what we call kleptoparasites," says Ng. "They'll often steal food from other birds." Despite what Franklin might have thought, however, being a klepto is nothing to be ashamed of (as an animal, at least; nation-states are another matter).

In the wild, animals do whatever they have to do to survive. Hyenas steal from lions. Lions steal from hyenas. Bears steal from squirrels. Bears steal from bees. Bees steal from each other. And on and on. You might even say kleptoparasitism makes the world go round.

And hey, did you know that the word "raptor"—which is often used to describe birds of prey—comes from the Latin verb *rapio,* which means to plunder, rob, ravish, or abduct? So yeah, bald eagles are like pirates, swooping in and taking what they want, when they want it. But let's not pretend doing so carries any sort of moral weight.

When they aren't thieving, bald eagles also scavenge from roadkill or help themselves to free food found in human garbage or at the town landfill. Again, none of this really means anything. You can know all of this and still love bald eagles and slap them all over T-shirts and bumper stickers and whatever else.

By the way, Ng confirms that bald eagle infatuation is very much an American enterprise. Folks in Canada and Mexico just see bald eagles as another bird, however impressive. But no matter where you're from, there's another piece of bald eagle lore people are starting to forget—that just a few decades ago, we nearly lost the birds to extinction.

The Bald Eagle's Comeback Story

Before Europeans colonized North America, bald eagles were pretty much everywhere there was a river, lake, or stream large enough to support sizable fish. In fact, one written account from New England in 1668 said bald eagles were so plentiful as to be "infinite." So much so, the colonists sometimes fed them to their pigs.

By the way, the bald eagle's common name hails from around the same time period, and it doesn't mean lack of hair. Rather, "bald" comes from the old English word "piebald," which is still used today for horses, and means coloration of alternating dark and light. As in, the bald eagle's dark brown body feathers versus its stark white head feathers. Oh, and those white feathers usually don't fully come in until the birds reach their fifth year of age, so young eagles aren't bald in either sense of the word.

Estimates vary and usually only include the United States, but during the colonial period, there would have been at least 50,000 breeding pairs and maybe up to as many as half a million bald eagles fluttering through the continent's skies. Of course, this is also where things take a turn, because lots of folks just weren't that fond of the birds.

"Eagles get a bad rap sometimes, and we see this in lots of parts of the world," says Ng. "Bald eagles scavenge on carrion, and so, if somebody goes out and they see a dead cow or a dead lamb and there's a bald eagle sitting on top of it, then they immediately think the bird killed it."

Similarly, in 1917, citing too much competition for their salmon harvest, the state of Alaska even instituted a bounty system that paid people to just go out and kill as many bald eagles as possible. A single eagle fetched two dollars from the state government, or at today's prices, nearly $50. At the same time, bald eagles were rumored to sometimes kidnap human babies and fly away with them in their talons. Throw in a feathered-hat fashion trend that swept the United States following the American Civil War, and frankly there were just too many incentives to go out and kill bald eagles. More than 120,000 eagles were slaughtered over the course of Alaska's bounty system alone.

All of this would have been detrimental enough to the continental bald eagle population. But then, we started killing bald eagles even more efficiently, and entirely accidentally. When America entered World War II, wartime manufacturing instigated a shift from the United States' main insecticide at the time, a naturally occurring compound known as pyrethrum, to another: dichlorodiphenyltrichloroethane, or as we've come to know it, DDT.

In the years following WWII, DDT overtook pyrethrum as the most popular standalone insecticide, and it worked well at its task—killing insects and other arthropods by disrupting their nervous systems, all the while seemingly not hurting people. Of course, today we know that high-dose exposures to DDT can cause all kinds of illness in humans, including vomiting and seizures. It's also a possible carcinogen. But back then, it was thought to be harmless to mammals. "There are pictures and videos of kids playing in plumes of DDT that the trucks would be releasing as they drove down the neighborhood roads," says Ng.

Of course, another problem was silently brewing. "The trouble with DDT is that it stays in the environment, and it gets washed away into water, and then plants, and little animals like insects and fish eat those little bits of residue," say Ng. Then, other fish eat those animals, and other animals eat those fish, and the pesticide just keeps concentrating as it moves up the food chain until, eventually, some of those poison-laced fish get eaten by a bald eagle (scientists call this process bioaccumulation).

Curiously, DDT did not kill the birds outright, but rather caused their eggshells to collapse under the weight of the incubating adults. Fortunately, says Ng, scientists studying these birds noticed that hatch rates were flatlining, so even though bald eagles didn't drop out of the sky en masse, scientists were able to identify the problem before it was too late. In 1972, the U.S. banned DDT under most circumstances. Canada followed suit beginning in 1985, as did Mexico in 2000.

It's taken some time, but bald eagle populations have slowly bounced back from a low of just 417 nesting pairs anywhere in the U.S. in 1963 to around 316,700 bald eagles in the lower 48 states. And scientists estimate there are at least another 100,000 nesting pairs in Canada and Alaska.

The comeback has worked so well, Ng says lots of folks she encounters in Canada now don't even realize the birds nearly disappeared. There's a scientific phrase for that, too. It's known as shifting baseline syndrome, and it refers to how we perceive what we see in the natural world—smaller fish, fewer insects, lack of large predators—to be the way it's always been. But the bald eagles and their return shows that our baselines can also shift in the other direction.

So if you should be so lucky as to see a bald eagle harassing an osprey for a fish someday, or maybe tucking in to a deer carcass on the side of the road, take a moment to reflect. We nearly lost these massive, terrifying, sort-of-silly gigglers forever.

GATEKEEPER FOR THE HEAVENS

CALIFORNIA CONDOR

SPECIES: *Gymnogyps californianus* **STATUS:** Critically Endangered **RANGE:** Pockets of the American Southwest and Baja Mexico **SIZE:** Up to 53 inches long, 10-foot wingspan; weighs up to 25 pounds **LIFESPAN:** Up to 60 years

---

FIFTEEN THOUSAND FEET SKYWARD, dark wings glide effortlessly in overlapping circles as the California condor searches for the telltale signs of death.

That far up, the sense of smell is mostly useless, so unlike other vultures, these birds rely on vision. A lump of rotting flesh, a flush of much smaller turkey vultures, or the flash of an eagle's wing—these are the calling cards the ancient scavengers use to identify their next meal. And once that target is locked, look out.

With a wingspan of nearly 10 feet and a body mass twice the size of the average supermarket turkey, the California condor is the largest land bird left in North America. And when they fall out of the sky and onto a carcass, every other wingèd thing in the vicinity knows it's time to skedaddle.

Vultures sometimes get a bad rap for being cowardly, since they tend to move in only after a prey animal has died. But when condors come calling, even golden eagles vacate the buffet within minutes. Of course, it probably helps that an army of up to 50 condors can feed on a carcass at the same time, and if they don't like who's there when they show up, the birds will open their beaks in a silent scream and charge at their rivals like red-eyed mimes.

But even as these gargantuan birds may act like bullies on a dead deer, they are actually givers, through and through.

"One of the neat things about condors is that, as the largest avian scavenger in the area, they can actually open up food resources that smaller scavengers, such as turkey vultures, can't access," says Tiana Williams-Claussen, Wildlife Department director for the Yurok Tribe, of which she is also a member.

Condors possess powerful feet that allow them to hold a carcass in place, while necks like pythons and sharp, curved beaks wrench their way through even the thickest of hides. Once inside, each bird can scarf down around four pounds of flesh and innards, which would be like one of us trying to stomach around 20 pounds of overripe steak tartare. Following a really good feast, the birds become so engorged that they sometimes have to wait several hours before they're physically capable of flying again.

Talk about a food coma.

After the condors have had their fill, dozens of other species move in, from turkey vultures and eagles to crows, coyotes, and skunks. Next, a parade of invertebrates that specialize in decomposition, such as carrion beetles and blowflies, arrives to finish off the carcass.

You and I are lucky enough to not have to think about it much, but the sooner this process begins, the sooner it ends. Without condors or some other apex scavenger to puncture a dead horse, cow, or seal, those bodies simply sit and rot. Bacteria are already inside, of course, and as they break down the banquet at their own pace, they release waste products, which include gases that bloat the carcass. At the same time, a whole host of noxious microbes you'd rather not have over for dinner begin to multiply. Nasty stuff, like anthrax and botulism.

Interestingly, when vultures gulp down meat with those kinds of microbes growing on it, not only do the birds not get sick, but most of the disease-causing nasties fail to survive the journey through their digestive tracts. Powerful acids in the vultures' stomachs attack the microbes' cell walls, while friendly bacteria (to vultures, at least) gobble up what survives to the digestive tract.

To give you a sense of how valuable scavengers can be, you need look no further than India at the turn of the century, says Williams-Claussen.

From the late 1990s to the early 2000s, India's three species of vultures experienced a mass die-off so sudden and dramatic that it took scientists years to figure out the cause. We're talking declines of 96.8 to 99.9 percent of the total population—nearly 40 million vultures dead in just 15 years. Experts later traced the plague back to the use of an inexpensive anti-inflammatory drug called diclofenac, which was given to cattle to reduce fevers and relieve pain. However, in vultures that came into contact with the drug after the cattle died, diclofenac caused renal failure, gout in the guts, and ultimately, a grisly death.

The whole saga would come to be known as the Indian Vulture Crisis. But what concerns us here is what happened next.

"There were all kinds of issues," says Williams-Claussen. "You had dead carcasses just sitting everywhere, which was very unpleasant, but it also led to a lot more disease. And not only diseases associated with the carcasses themselves."

Without vultures to break down the bodies, feral dogs moved into the vacuum. In time, the dog population exploded, which in turn correlated with an increase in rabies cases among humans. Between 1992 and 2006, an extra 38.5 million rabies-laced dog bites are thought to have killed 47,300 people in India.

"So it was just a gnarly mess," says Williams-Claussen. "And that is the service that these kinds of birds are providing."

An Apex Scavenger Returns

As with India's vultures, the California condor population has undergone a free fall. Today, slightly more than 560 of the birds remain, with around half of those living in captivity.

The condor's decline seems to have begun when Europeans first started colonizing North America. Not only did a rapid increase in humans and their gun-enabled hunting mean fewer game animals everywhere, but settlers tended to be antipredator, and the cyanide and strychnine they laced carcasses with to kill wolves, mountain lions, coyotes, and bears ended up annihilating scavengers like the condor, too.

Lead ammunition is a poison all to itself. When a bullet strikes a piece of meat, it can splinter and send shrapnel in every direction throughout the body. And when animals hit by lead bullets die and condors eventually find their way to their carcasses, the birds scarf down those neurotoxin nuggets without even knowing they're there.

It doesn't take much to kill, either. Ingesting a morsel of lead the size of your fingernail is enough to drop a bald eagle. Unfortunately, the problem is not one of the distant past. Around half of all condor deaths between 1992 and 2020 were due to lead.

The Yurok call the condor *prey-go-neesh,* and in their culture, the animal stands as a gateway between worlds. Condors fly higher than almost any other bird, and for that reason, the Yurok believe that it's this animal who delivers their prayers to the heavens.

"Our connection to the condor goes back to what we consider to be the beginning of time," says Williams-Claussen.

According to Yurok tradition, the Creator designated this tribe, as well as others in the region, to be what's known as world renewers, or people whose purpose is to help fix the planet, Williams-Claussen says. And part of that duty is to perform world-renewal ceremonies each year. While such events are private and sacred affairs for her people, Williams-Claussen shared that prey-go-neesh features prominently in the 10-day rites.

One of the ways Williams-Claussen and her people fulfill that world-renewal role today is by bringing the California condor back from the brink of extinction.

After the birds' population crashed down to fewer than 35 individuals in 1979, all the remaining birds in the wild were rounded up and brought into captivity. There, experts cradled the species and protected it, simultaneously raising and releasing as many chicks as possible. But there are still many places where condors used to soar that haven't seen their silhouette against the sun in a hundred years or more.

The Klamath River watershed, including where the Yurok still live today, is one such place. Or it used to be.

In 2022, Williams-Claussen and her team brought four fledgling condors into a facility at Redwood National Park where they were nurtured and allowed to get to know each other. Then, one by one, the birds were released into the clear blue sky.

The birds aren't all on their own just yet. The Yurok still monitor the animals daily by radio-tag, offer supplemental food—often organic dairy cattle calves that were stillborn and volunteered by local farms—and even capture the birds at least twice each year to test their blood levels for lead, an ever present threat, as well as emerging diseases, such as the highly pathogenic avian influenza.

The hope is that the Yurok Tribe will continue to release four to six birds each year for the next 20 years. Condors are highly social animals, so a large flock where the birds could regain part of that shared responsibility to find food would be ideal. The birds also pair for life, says Williams-Claussen, so taking the long view makes the most sense for the future of the species, as well as her people.

"There's a lot of emotion that is bound up in having a bird in your arms," says Williams-Claussen, who is one of the few people on Earth to ever hold a California condor chick. "You see its whole future ahead of it. It's not that different from holding on to your toddler and worrying about their future."

At least one thing remains certain. So long as Williams-Claussen has anything to do with it, her daughter will be among the first generation of Yurok children in a very long time to grow up in a world with condors overhead. World renewal, indeed.

THE SOFT-WINGED SENTINEL

MEXICAN SPOTTED OWL

SPECIES: *Strix occidentalis lucida* **STATUS:** Threatened **RANGE:** Southwestern U.S. and northern Mexico **SIZE:** Up to 19 inches long, wingspan of 45 inches; weighs up to 23 ounces **LIFESPAN:** Up to 15 years

---

AT AN UNDISCLOSED LOCATION in the Four Corners region of the U.S., where the southern Rocky Mountains bristle with Douglas and white fir trees, Serra Hoagland walks through the pines at dusk doing her best impression of a Mexican spotted owl. To my ears, the owl's four-note-hoot sounds a bit like a half-asleep Pomeranian yapping at a car on the street. But this is only one of the vocalizations these dark-eyed birds produce.

"They also have a series of really interesting barks or whistles," says Hoagland, who is a biologist with the U.S. Forest Service and tribal member of the Laguna Pueblo. "And even the young, they have this very cute cooing and almost a purring. It's almost like if you would imagine a little kitten is up in the forest, way high up in the trees."

North America is home to three subspecies of spotted owl, including the California spotted owl and northern spotted owl. The latter is probably the best known of the bunch, because that subspecies was the subject of a highly public debate between environmentalists, who thought the animal should be protected, and loggers, who argued that doing so would put them out of work. Tensions swelled, death threats were issued, and the northern spotted owl even landed on the cover of *Time* magazine. Ultimately, the U.S. Fish and Wildlife Service added it to the endangered species list in 1990, though the ruling remains controversial to this day.

Most of the year, Mexican spotted owls live solitary lives, haunting old-growth forest and canyonlands as they hunt for small birds, bats, reptiles, and bugs. To kill their supper, the owls perch high above their prey, using the scritches and scratches of movement below to zero in on and dive-bomb their meal.

One of the things that allows owls to use sound as a targeting system is a set of asymmetrical ears. You can't see the spotted owl's ears because they are hidden behind their feathers. But beneath the fluff, spotted owls have holes that collect sound, just like you and me. And because one of those holes is slightly higher than the other, sound waves hit them at slightly different times. In some owl species, the earholes may also be slightly closer to the face or toward the back of the bird's head. Whatever the precise arrangement, asymmetrical ear openings allow the predators to triangulate the chitter of a woodrat or the squeak of a bat and then use that information to determine where to strike.

Of course, Mexican spotted owls most often hunt after dark, so night vision is another critical asset. And just like their ears are a little weird, well, so are their eyeballs. Because they're not really balls at all.

Unlike humans, owl eyes are more tube-shaped than spherical, and scientists believe this may have been an evolutionary workaround that allowed the birds to develop large eye organs despite having rather tiny skulls. And when I say *large* eyes, I actually mean *humongous*. Because owl eyes make up 5 percent of their entire body weight. In humans, eyes account for just 0.0003 percent.

Owl eyes are so big that you can actually see the backs of them by peeking inside the birds' earholes. Those massive eyes are stuffed with rods, the eye cells that detect light and movement, especially in the dark. While owl eyes are binocular like those of humans, allowing the birds to perceive depth, owls also have a special *tapetum lucidum,* or a layer behind the retina that reflects low levels of light back through the eye's receptors for a second time. This quirk of biology provides the owl's eyes with a double dose of whatever light is available, the effect of which is straight-up night vision that can be as much as 100 times more sensitive to low light than human vision. (The tapetum lucidum is also what makes an animal's eyes glow when caught in the beam of a flashlight.)

Of course, all of that visual machinery comes at a cost. As owl eyes filled their skulls, they lost the ability to move in their sockets. This is why an owl must turn its head to look in a different direction. And while the birds make up for that inadequacy with the ability to rotate their necks an impressive 270 degrees, this created a different problem—twisting their necks that much actually kinks up the blood flow from their heart. Incredibly, the birds evolved *another* workaround for this, too. Oxygen-rich blood actually pools in their brains and eyes so that if the animal moves its head and cuts off circulation, the important bits still get what they need.

Come early spring, male and female Mexican spotted owls meet up to mate, usually with the same bird they shacked up with in years prior. While Mexican spotted owls share parenting duties, tasks are strictly delineated. Usually, it's Mom who incubates the eggs, while Dad hunts for food. And this actually aids Hoagland in her mission to spot owls amid the woodwork. After all, a single breeding pair might lord over a core area as large as 100 acres, with another 500 acres serving as their foraging grounds, says Hoagland, who has been working with the subspecies for more than a decade.

"They have this territorial behavior about them during the nesting season. They protect their home ranges through the vocalizations," she says. In other words, when Hoagland hoots, the pair feels threatened and quickly hoots back, which tells her there's a Mexican spotted owl watching her from somewhere above. Of course, humans can't triangulate sounds quite as efficiently as owls, so she may be able to hear the pair but not see them. And that's when Hoagland will deploy another trick of the trade—a large white feeder mouse from a pet store.

Nesting Mexican spotted owls cannot resist an easy meal. So if Dad's around, he will materialize out of the forest canopy, talons flared, sometimes so quickly, Hoagland hasn't even fully stepped away from the rodent. Then, with a few nearly silent wingbeats, the male will haul his prize back up into the treetops to feed Mom and whatever chicks have hatched.

"Woodrats are really like their cheeseburgers," says Hoagland of the owls' natural prey. "So we use that baiting method to identify exactly what tree is used as the nest tree."

Sounds a little cruel to the sacrificial mouse, right? Well, it's probably good to remember that these birds aren't doing so hot, as a species. While Mexican spotted owls have received much less attention than their cousins in the north, they, too, are birds in peril.

The Triple Bottom Line

Like northern spotted owls, the old-growth forests that Mexican spotted owls call home are increasingly targeted for timber operations. But there's another problem—fire.

"These places that they utilize for nesting habitat are also places that are susceptible to stand-replacing fire," says Hoagland, referring to the huge, catastrophic wildfires that now make the news with regularity. And this is all a result of more than a century's worth of suppressing naturally occurring fires, which were once considered a negative for the landscape. (Climate change is making all of this worse, too, by the way.)

These days, most forest managers see small, semifrequent fires as a healthy part of a balanced ecosystem. But rectifying a century's worth of mismanagement is neither easy nor cheap. Adding to this complex problem is the fact that large swaths of Native American lands lack the funding, infrastructure, and personnel to address these historical problems.

Fortunately, the science of fire ecology is pretty well understood these days, says Hoagland. The more difficult part is trying to figure out how to balance what's right for the forest, the owls and other endangered species, and the people who live in and depend on these lands. To bridge that gap, Hoagland says we need to take a more holistic approach to forest management—one that utilizes both Western science and Native perspectives. The approach is known as the Triple Bottom Line.

"It's really about promoting the social needs, the environmental needs, and the economic needs of that community," says Hoagland, who is also a Tribal Relations Specialist for the U.S. Forest Service's Rocky Mountain Research Station.

In practice, this means creating a more resilient forest through selective thinning and prescribed burns so that fuels can't build up over decades and then disappear in a landscape-scorching wall of flame. Opening up the canopy here and there is also good for Mexican spotted owls, says Hoagland, because it creates grassy areas where prey thrive. Not to mention all the other species living in the owl's shadow.

"The owl really is a good indicator species for the overall health of the community. The sites they are occupying also have a lot of bird species diversity and tree species diversity. They have really beautiful soils," says Hoagland. "It's a species that is emblematic of resilient and healthy forests."

On a personal note, Hoagland says there's just something about studying an animal that answers your call that she can't put into words.

Perhaps some sentiments can only be expressed in hoots.

THE ARROW OF THE GODS

RUBY-THROATED HUMMINGBIRD

SPECIES: *Archilochus colubris* **STATUS:** Least Concern **RANGE:** Eastern portions of Canada, U.S., Mexico, and Central America **SIZE:** Up to 3.5 inches long, wingspan of 4.3 inches; weighs up to 0.2 ounce **LIFESPAN:** Up to 9 years

---

IN THE BEGINNING OF THE WORLD, the Maya gods took a piece of shimmering jade and carved it into an arrowpoint. Then, with divine breath, the gods coaxed the tiny warhead to life and charged it with a sacred purpose—the delivery of all their secrets and desires. Today, we know this sentient thought-projectile as the hummingbird.

The Maya aren't specific about which hummingbird species served as messenger to the gods, but there are 330 species described by scientists to date—all of them found in the Americas, and the Americas only. It may have been the ruby-throated hummingbird, because though these fiery sprites are often spotted as far north and east as Canada's Newfoundland in the summertime, in the winter, the thumb-size, emerald-colored sprites migrate all the way through the heart of what used to be the Maya Empire in Mexico's Yucatan Peninsula. Some of the pipsqueaks have even been known to fly up to 22 hours nonstop over the Gulf of Mexico—a distance of more than 500 miles. Quite a feat given that hummingbirds are the smallest birds on Earth.

And yet, within a feathered frame that weighs less than a stick of chewing gum is a magnificent muscle larger than that of any other animal, relative to body size—the hummingbird heart. With superior blood-pumping power, some hummingbird species can beat their wings up to 90 times in a single second, which allows the animals to achieve flight patterns other birds can't. Things like flying sideways and backward, or hovering perfectly still in midair while licking nectar out of a flower with a tongue so long that it's anchored to the top of the birds' skulls. During courtship dives, some hummingbird species can approach speeds of nearly 60 miles an hour.

Hummingbird feathers glisten in the sun like wet rainbows, not because they're pigmented, but because they're stuffed full of iridescent melanosomes—microscopic, pancake-shaped structures that reflect light in so many different ways, we can't even see them all with our human eyeballs. Hummingbirds lay eggs the size of jelly beans and build nests out of lichen, moss, and spiderwebs. They are, without equivocation, one of the coolest and weirdest and most extreme living things on this entire planet, and as evidenced by the Maya lore recounted above, we are far from the first humans to marvel at their compact wonder.

In another story—this one belonging to the Aztec—the hummingbird fell to Earth as a ball of feathers. Cōātlīcue, the Earth goddess of fertility, nestled the creature to her breast and became pregnant with its spirit. This angered her sons and daughter, who either felt shamed or did not want another sibling, so they conspired to murder the goddess and her unborn child. However, before they could accomplish the deed, the hummingbird's spirit burst forth transformed. Gone was the dainty bird, and in its place stood a vibrantly adorned warrior known as Huītzilōpōchtli, god of war and the sun. With a flaming serpent for his weapon, Huītzilōpōchtli laid waste to his brothers and sister, the latter of whom was decapitated with such fury that her head flew into the sky and became the moon.

The Aztec also believed that the hummingbird god's war was ongoing. Each night, Huītzilōpōchtli did battle against his moon-sister and brothers (who became the stars), and each night he emerged victorious, as evidenced by the rising of the sun. The only problem? The hummingbird god's terrible power needed to be replenished daily, and there was only one way to do that—by offering him hearts ripped from the chests of captured enemy warriors and slaves.

Yes, the human sacrifices that have become emblematic of one of the ancient world's most formidable civilizations were carried out in the name of a *humming-*

bird. Or a hummingbird deity, at least. Stop feeding Huītzilōpōchtli hearts, the Aztecs believed, and he would fall in battle and the world would plunge into forever-night.

"Nowadays, we tend to see hummingbirds as small and delicate creatures," says Vanessa Hernández Urraca, a doctor in veterinary medicine who has written about hummingbirds and Mexican culture. "But our ancestors saw braveness and ferocity."

Turns out, the ancestors had it right.

Flower Power

A flash of feathers, a buzz of wings, and a duel of needle-nosed bills—watch a hummingbird feeder long enough, and you may just catch a glimpse of what it looks like when fairies feud.

On rare occasions, hummingbird tussles can even leave losers impaled by the victor's stiletto. Sometimes lethally so. So why do these tiny terrors seem to throw down more than other birds?

Well, a hummingbird lives its life in overdrive, floating this way and that, and generally defying the laws of physics. And to do so, these birds must consume nature's rocket fuel: nectar. Depending on the plant that produces it, nectar can contain up to five times the sugar content of a can of Coca-Cola. The catch, of course, is that each flower only has an itty bitty nip of nectar lurking within. So to keep their energy budgets in the black, hummingbirds might have to wet their beaks at more than a thousand flowers every single day. And getting from flower to flower requires that they burn oodles of calories, so every hour these birds are yo-yoing between peaks and valleys of energy as they add and subtract nutrients (snagging a few insects, spiders, and laps of tree sap now and again also helps rubies meet their needs). Overall, while adult humans need around 2,500 calories each day, if hummingbirds were the same size as us, they'd need around 10,000 calories each day just to do what they do.

At night, things get even more extreme for the hummingbird. Because a hummingbird can't afford to exist—to breathe, to digest, to stay warm, to pump blood—without constantly adding gas to its tank, it has evolved the ability to turn the engine down so low, it's nearly off.

"They reduce their temperature and their metabolism to survive the night when there's no food," says Hernández-Urraca. Scientists call this nightly power-down torpor, but you can think of it like a mini-hibernation. And it is probably what led to another legend that isn't so far off from reality. According to Hernández Urraca, it was commonly believed that hummingbirds died every night only to be reborn each morning. Sort of reminds you of a certain snake-swinging god, right?

Suffice it to say that every day in the life of a hummingbird is spent raging against the dying of the light, and that means they cannot suffer another hummingbird encroaching on their private stash of life-sustaining flowers. At the same time, intruders are just trying to survive themselves, so they also have motivation to fight.

"Hummingbirds can pollinate more than a thousand different flower species," says Hernández-Urraca, "but one medium-size feeder is the equivalent to 2,500 flowers." This means the feeder outside your window isn't just a few glugs of sugar water, it's a hummingbird holy grail. And sometimes, when natural nectar sources are scarce, a feeder can be worth fighting to the death over.

By the way, if you want to reduce violence among your backyard birds, set out several smaller hummingbird feeders rather than one big one. Also, clean your feeders often and skip the red food coloring, says Hernández Urraca. Otherwise you might be exposing your hummingbirds to pathogens and additives.

Food is just one reason hummers fight. Males have evolved brilliant color displays that reflect sunlight from different angles, and these are used to both advertise themselves to the ladies and also warn off challengers. For instance, male ruby-throated hummingbirds are easy to spot because they have stunning red patches of feathers on their necks, known as gorgets, while females sport white or gray neck feathers. And if the male's visual deterrent isn't doing the trick, then sometimes standoffs come to blows.

Females are also capable of exponentially more wrath than you might expect, though. After mating, male rubies buzz off to try their luck with another bird while the females must build a nest and then defend it from predators, including much larger species such as crows or raptors, says Hernández Urraca.

In some cases, hummingbirds have even been seen attacking golden eagles, which outweigh hummingbirds by about a thousand times. That isn't exactly the same as Luke Skywalker taking on the Death Star, but you get the idea.

Any way you slice it, ruby-throated hummingbirds are not the prim and prissy creatures we've sometimes painted them to be. They are balls of fury descended from the clouds of myth and legend. Each is a blur, a wisp, a cluster of electrons filed down to a knife's edge by the gods and polished by evolution until all that remains is sugar-starved rage and iridescent wonder.

So put out your feeders, plant native flowers, and tip your hat to the hummingbird god himself, because according to the Aztec, we'd all be living in darkness without these needle-nosed nectar fiends.

Northwestern salamander
(Ambystoma gracile)

PART III

REMARKABLE REPTILES & AMPHIBIANS

THE ARMORED SUBMARINE

AMERICAN ALLIGATOR

TYPE: Reptile **SPECIES:** *Alligator mississippiensis* **STATUS:** Least Concern **RANGE:** American Southeast **SIZE:** Up to 14 feet 9 inches long; weighs more than 1,000 pounds **LIFESPAN:** Up to 80 years

---

HAVE YOU EVER THOUGHT about how weird it is that there's an armored reptile on this continent that can weigh as much as a Thoroughbred horse? Or that these cousins to the dinosaurs possess one of the hardest-hitting bites of any living creature on the planet—stronger than hyenas, hippos, and jaguars? A chomp so powerful, it feels like being pinned beneath the weight of a pickup truck?

Now, can you imagine willingly placing your arm inside the mouth of one of these creatures for science?

We can learn a lot about an animal by studying what's in its stomach, and fortunately, there's a relatively easy way to do this with alligators. It's called the hose-Heimlich method, and it's basically like pumping the giant reptiles' stomachs. Though effective for research and safe for the gators, the method does have its risks for the scientists.

Adam Rosenblatt, an ecologist at the University of North Florida, remembers one time when he was up to his armpit inside another crocodilian—the black caiman in Guyana—when tragedy nearly struck.

"I put a PVC pipe into the mouth of an 11-footer. The pipe keeps the mouth open while we're pumping the stomach and allows us to put our hands in the caiman's throat to pull out any large prey items that might get stuck," he explains. "At one point I put my hand in the caiman's mouth to dislodge a piece of a crab, and just as I took my hand out, the caiman's jaws broke through the PVC pipe and crushed it."

He paused. "Now we only use metal pipes."

Though the animals are obviously capable of inflicting great harm, most of the time they don't want anything to do with us. "They don't see us as prey, especially when we're adults," says Rosenblatt. "Think about it from the alligator's perspective. Even a 10-foot-long alligator, they're very low to the ground, right? Their body is horizontally long, not vertically long. Whereas, as adults, we are five to six feet off the ground. And so we tower over them, and it's very intimidating for them."

Not that alligator-on-human tragedies never happen. They do. But between 1948 and 2021, just 26 people died in the grips of an alligator. Fireworks kill people more frequently, for what it's worth.

By and large, almost all alligator attacks come down to just a few factors, says Rosenblatt. First, you have people accidentally stepping on the animals when they're in an alligator habitat, which might sound ridiculous, but let's remember that alligators' dark, knobby skin makes for excellent camouflage. Also, the reptiles can stay underwater for up to two hours at a time. And there's evidence they actually swallow rocks as ballast to extend their dive times. So you truly might have no clue there's an alligator anywhere near you.

Second, female alligators will fiercely defend their nests, something Rosenblatt has witnessed while trying to collect eggs for his research. So if people accidentally wander too near, the mama gator might get feisty. It's kind of cool actually, considering that we often think of reptiles as being cold-blooded, noncaring critters. But with alligators, Mom actually sticks around and listens for chirps from her eggs, at which point she begins to remove nesting material. This makes it easier for the babies to hatch and get to the safety of the water. Mom might even protect the babies for several years, which is necessary because baby gators can fall prey to everything from opossums to wading birds to bass.

The third scenario that gets us humans into trouble is when people feed wild alligators. Doing so makes the animals associate humans with snacks, which scientists call food habituation. Rosenblatt says it's by far the most common situation that leads to gator problems. Because an alligator looking for handouts can become violent when it doesn't receive them.

Finally, there's the problem of pets. While even the largest of gators won't look at adult humans and hear a dinner bell, our kitty cats and cocker spaniels are another story entirely.

"Alligators are the classic opportunistic-generalist predator," says Rosenblatt. Put another way, alligators are armored submarines equipped with infinite patience and a knockout punch. And that arrangement allows them to devour nearly 70 different families of animals, according to Rosenblatt's research. "The basic rule of thumb is they'll eat anything that they can fit their mouth around." And while those num-nums can include big prey, like deer and pigs, which they latch onto and then drown, most of the time alligators seem to order appetizers over entrees.

"People are always shocked when I do stomach contents work on alligators," says Rosenblatt. "What comes out of almost every alligator is only really small stuff. Even for the largest, 10- to 11-foot-long alligators, you're getting pieces of crabs, you're getting pieces of turtles, you're getting fish, you're getting shrimp, you're getting insects. Like, they are fundamentally lazy animals. And I mean that in the best way possible. They go for the easiest meal. And if that means they're eating snails, then they're eating snails."

Even if alligators are generally super chill, if you're in the American South and thinking about going for a swim in a lake or river, the safest thing you can do is go ahead and assume that there's a giant reptile nearby. It's probably not true, says Rosenblatt, but doing so could keep you safe.

"It depends on your personal level of risk tolerance, obviously, but I would not put my children in a natural, fresh body of water in Florida for swimming unless I knew that body of water really well," he says.

There are also two things you should know in the unlikely scenario that an alligator attacks. The first is that the old adage about running away from a gator in a zigzag pattern? "That is just ridiculous," says Rosenblatt. These reptiles aren't built for sprints on land, and they rarely ever chase prey there. Alligators move at a top speed of 10 miles an hour in water and are slightly slower over short distances on land. In other words, in the unlikely event that you have a gator on your heels, just run as straight and fast as your legs can carry you.

The other? While we've already talked about just how powerful an alligator's bite is, the muscular mechanism that powers it is only mighty in one direction. "They have almost no opening power," says Rosenblatt. "So if you can get the jaws closed, you could hold them closed with one finger applying not a huge amount of pressure."

Finally, if a gator ever has you in its grips, know that there is only one vulnerable part of an alligator's head. "Your best shot at escaping is to jam a finger into one of its eyes," says Rosenblatt.

The Saving Grace of Alligator Skin

As with white-tailed deer, bald eagles, American buffalo, and too many other species, the fact that we still have American alligators around today, and in such abundance, was not always a given. In fact, the reptiles were one of the first species added to the United States' endangered species list in 1967.

As for what drove alligators to the brink of extinction, well, it was that beautiful skin. Products made from tanned alligator hides became all the rage after the American Civil War, leading to an all-out slaughter that did not abate until state and federal governments united to provide restrictions on poaching and exploitation. "We're talking about, like, a hundred years of intensive hunting," says Rosenblatt. Between 1880 and 1955, people in Florida and Louisiana alone were estimated to have harvested six million of the leathery reptiles.

Are you ready for the twist? Even though it became illegal to kill alligators in the wild, there was nothing stopping people from taking eggs from the wild and then essentially farming the animals in captivity. "So what they did is they basically flooded the market with high-quality, cheap skins," says Rosenblatt. "And that led to a decline in the value of skins overall, so there was much less financial incentive for poachers to be illegally taking animals from the wild."

At the same time, farms in Louisiana were actually boosting the overall population, because by law, 12 percent of the alligators raised had to be returned to the wild. And the reason this was huge, says Rosenblatt, is because gators have a very low survival rate in the wild in those first few years. But the farmers were putting their thumbs on the scale, in essence, by taking the eggs, raising them in safety, and then returning gators back into the bayou at sizes that kept them safe from most predators. So the alligator's armor was both its blessing and its curse, and then its blessing again.

In 1987, just 20 years after the U.S. Fish and Wildlife Service added the American alligator to the endangered species list, it removed it.

Alligators went from the edge of extinction to the point where we might now assume one could be cruising through any body of fresh water in the American South. That's about as unlikely a turnaround as you'll ever hear about, and not one enjoyed by many other thousand-pound predators.

THE GILL-FACED GOD MONSTER

AXOLOTL

TYPE: Amphibian **SPECIES:** *Ambystoma mexicanum* **STATUS:** Critically Endangered **RANGE:** Mexican Central Valley **SIZE:** Up to 12 inches long; weighs up to 8 ounces **LIFESPAN:** Up to 10 years

---

IN THE CANALS OUTSIDE OF MEXICO CITY, there lingers a large, banana-size amphibian known as the axolotl. Named after the ancient Aztec word for "water monster," the axolotl is unlike most other creatures on this planet, in that these giant salamanders possess the nearly magical power of regeneration.

Gills, feet, legs, tails, hearts, even pieces of the axolotl's own spinal cord and brain—you name it, and the axo can rebuild it.

"If it loses an *eyeball,* it can grow it back," says Luis Zambrano González, an ecologist at the National Autonomous University of Mexico.

In some instances, axolotls have even shown that they can regenerate the same limb up to five times in a row without suffering any negative effects. That's a superpower more commonly found in the pages of a comic book than the annals of real life. All of which begs two huge questions: Why did axolotls evolve the ability to rebuild themselves? And uh, can they please teach us how to perform the same miracle?

As to the former, some suspect the axolotl's trick may help the amphibians survive their early years, which are spent around many other axolotls. This is because axolotls have a tendency to nip each other's gills and limbs off.

But regeneration capabilities could also just be a workaround for an animal that doesn't have much fight in it. "They are very peaceful and calm," says Zambrano. "They don't run when they see a predator." Instead, they hunker down and hope their camouflage cloaks them against the murky water and tangles of vegetation they inhabit. Better to lose a leg that can be regrown than waste precious energy trying to fight or flee.

And this brings us to an important point. While axolotls have become a popular animal in recent years, suddenly turning up in kids' cartoons and as plush toys, nearly all popular depictions of this species portray it as bright white or pink. And it's true, the animals do possess a combination of pigmentation genes that, in rare instances, can produce axolotls with bubble-gum-pink or ghostly white skin. But such creatures would be picked off almost instantly by predators, says Zambrano, which is why you don't find ivory- or coral-colored salamanders in wetlands outside of Mexico City—the only place axolotls live. Rather, the wild axolotl comes in an array of mossy greens, muck browns, and midnight blacks, the better to disappear.

So where did the idea that axolotls are pink and white come from? Well, scientists have been keeping these animals in captivity since 1864. And we've learned quite a deal from the amphibians in all that time.

For instance, we now know that the axolotl's genome is more than 10 times as long as a human's, says Zambrano. Nestled within that voluminous spool of code are a whole bunch of genes that give axolotls their renewal abilities—scientists are already targeting those, using axolotls as a model organism that can be used to help us understand how regeneration works. But the species is also spectacularly resistant to cancer, a trait we very much want to borrow.

What's more, axolotls approach life with a Peter Pan sort of vibe. They literally never grow up but rather spend their entire lives as juveniles. The headdress of feather-like gills ringing the axolotl's face is a giveaway of this perpetual teenager state. Most other salamander species lop those gills off as they transform into land-living adults. Axolotls, on the other hand, spend their whole lives underwater. Scientists call this bizarre state of prolonged adolescence neoteny, and they suspect it might have something to do with the axolotl's gift of supercharged healing.

Anyway, with so much medical relevance, scientists perfected how to keep captive populations in labs all over the world. And that, in turn, led to the animals becoming darlings of the aquatic pet trade, which is well-known for creating unusual color morphs and patterns through selective breeding. At the same time, Zambrano says there was a scientific incentive to breed lab animals with lighter coloration because it allowed researchers to more easily see inside the salamanders as they tagged various cell lines with green fluorescent protein. (Derived from an Australian jellyfish in the late 1990s, green fluorescent protein has become an incredibly useful tool for scientists studying how human bodies work.)

Unfortunately, even with their mystifying regenerative abilities and enormous captive population, by the time you read this, these forever-young oddballs may well have already gone extinct in the wild.

Axolotl-Safe Salsa

In 1998, scientists estimated around 6,000 axolotls remained in the canal system of Lake Xochimilco just south of Mexico City. In 2004, that number dropped to 1,000 animals. By 2008, the population of wild axolotls was thought to hover somewhere around 100. "Now, I'm sure it is less than that," says Zambrano, one of the world's few scientists who studies the axolotl in its rapidly deteriorating and shrinking habitat.

Oh, but what a range it used to have. You see, in most freshwater lakes, life gravitates to the edges—places where tree roots, cattails, and other aquatic vegetation provide hiding spots for small fish and invertebrates. "Very few species of animals and plants exist in the center," says Zambrano. But Lake Xochimilco hasn't been an ordinary lake for a very long time.

For the past thousand years, several societies have risen and fallen on the shores of this body of water, starting with the Xochimilca people and leading into the reign of the Aztec. But while the surrounding people and culture changed through the years, one thing remained steadfast—a commitment to building chinampas, or floating gardens.

Lake Xochimilco is only a few feet deep and, on its bottom, blanketed in rich, black sediment. Early inhabitants of the area learned that they could dredge up that underlying fertility and slather it across a matrix of willow branches and stones in layers, kind of like a cake. Over time, these floating flower beds would falter and sink, and people constructed new chinampas on top of them, to the point where the structures eventually became permanent, artificial, hydroponic islands. The agricultural output of the chinampas was so powerful that the Aztec used them to basically feed the entire city of Tenochtitlan, which at one point was the largest urban center in the Western Hemisphere and would later serve as the base out of which Mexico City grew.

All of this infrastructure was good for axolotls, because each chinampa was surrounded by a grid of interlocking canals that maximized the lake-edge habitat. Instead of one shoreline, the salamanders benefited from thousands of edges where they could hide away from predators, lay their eggs, and hunt crustaceans, mollusks, insect eggs, and small fish. Conditions led to the proliferation of so many of the frill-faced amphibians that axolotl tamales were considered a local delicacy all the way up until just a few years ago, says Zambrano.

Alas, the axolotl paradise would not last forever. You could say the downfall began in 1521 when the Spanish conquistadores laid siege to Tenochtitlan for three months, during which smallpox seethed within the city limits. Eventually, the Aztec would succumb, and the colonizers would remake the city in their own image—a transformation that would see wetlands drained and dried to make way for more European styles of agriculture. This reduced the axolotl's habitat considerably, says Zambrano, but the final hit came in the 1950s, '60s, and '70s when what remained of the chinampas-farming lifestyle fell out of favor in and around the city.

In recent decades, urbanization has claimed most chinampas and converted them into everything from housing to soccer pitches. Just around 2.5 percent of the floating gardens remain functional today, says Zambrano. Pollution has skyrocketed, souring the canals with trash, industrial fertilizers, and pesticides. At the same time, invasive populations of tilapia and carp—introduced in the 1990s as a source of food for people—prowl the corridors, making meals out of axolotl eggs and young. The International Union for Conservation of Nature lists the axolotl as critically endangered, one step away from extinction.

But Zambrano and his colleagues aren't giving up on the axos without a fight. In recent years, they've experimented with outfitting some canals with filters that both improve the water quality of the area and also keep out salamander-munching fish. There's also been a push to reinvigorate the chinampas culture and the improved water quality and axo habitat that comes with it. We already know that chinampas can churn out high-value crops, such as chiles, corn, herbs, tomatoes, and beans and can do so without pesticides or artificial fertilizers. And if fancy restaurants in Mexico City source their produce from chinampas, the hope is that products labeled axolotl-safe will fetch a premium price point. Hey, it worked with dolphin-safe tuna, right?

So far, these efforts have restored around 40 chinampas and produced about three miles of axolotl refuge. But if the species is to be saved, he says, everything needs to move more quickly. "We are in this race against time, and we are losing," says Zambrano.

The good news is the scientists do still have one axolotl in the hole—the vast, easy-to-breed captive population. The catch is that even if you still have loads of axolotls, you can't reintroduce a species if it no longer has a habitat.

THE SNAKE FOR SNAKE HATERS

EASTERN INDIGO SNAKE

TYPE: Reptile **SPECIES:** *Drymarchon couperi* **STATUS:** Least Concern
RANGE: American Southeast **SIZE:** Up to 8 feet 7 inches long; weighs up to 11 pounds
LIFESPAN: ~26 years (in captivity)

IN THE LONGLEAF PINE FORESTS of the American Southeast, there slithers a veritable beauty queen of serpents. Scientists call it the eastern indigo snake.

With shiny black scales, an indigo snake in the sun may remind you of the inside of a piece of newly cracked coal. And that shimmery, inky black is accentuated with a patch of reddish-orange scales below the chin that pops and sizzles in a way that's just plain fun to look at. Best of all, in the right light, an iridescent sheen radiates across their bodies and makes the reptiles look as if rainbows are dancing across their coils.

While many folks have never even heard of this species—nonvenomous, by the way—indigos are among the largest snakes native to North America; they may even be *the* largest. To size up the typical adult eastern indigo, you'd need a measuring tape, because the suckers tend to be between five and seven feet in length. However, some eastern indigos have been measured at lengths of more than 8.5 feet. (There's some discussion as to whether the gopher snake might actually best the indigo in length, but nevertheless we're talking about a stunning serpentine specimen here.)

"This is a really special animal," says David Steen, leader of amphibian and reptile research at the Florida Fish and Wildlife Conservation Commission. "Even people who don't tend to like snakes often can appreciate indigos."

And there's one simple reason for that.

"They eat other snakes," says Steen.

Snake Eaters

There are actually a number of indigo snake species and subspecies, including the Texas indigo snake *(Drymarchon melanurus erebennus)* and the Middle American indigo snake *(Drymarchon melanurus).* But in the U.S. states of Florida, Georgia, Alabama, and Mississippi, the eastern indigo snake reigns supreme. Not that you're all that likely to stumble across one.

Despite being large and in charge in the longleaf pine forests they call home, eastern indigo snakes keep to themselves and spend a majority of their lives tucked away belowground—usually in a gopher tortoise burrow or some other subterranean lair. Because the eastern indigo cannot create its own burrow, the species' fate is tied in a knot with that of the burrow-building gopher tortoise. When the tortoises start to disappear due to habitat loss, so will go the indigos.

Now, if you're wondering how a nonvenomous snake regularly dines upon serpents that have fangs full of venom strong enough to kill a horse, well ...

"They'll often use brute strength," says Steen, the author of *Secrets of Snakes: The Science Beyond the Myths.* "They have these huge, powerful jaw muscles, and they will basically chew a rattlesnake's face to death."

While boas squeeze the life out of their prey, inducing cardiac arrest, and pit vipers rely on chemical weapons to send their enemies into physiological chaos, the indigo mauls its targets like a rabid dog. Whether lunch is a mouse, bird, frog, turtle, or even a small alligator, indigos seize their prey by the head and pin them to the ground or the wall of a burrow, gnawing as they overwhelm with superior strength and bite force. Sometimes, the prey is dazed but visibly still alive as it disappears down the serpent's maw.

A few years ago, Steen and some of his colleagues decided to test a few long-suspected theories surrounding these charismatic reptiles. For starters, the scientists wanted to know whether indigos actually preferred eating other snakes over other forms of prey.

The curiosity was more than academic. Indigos have disappeared from some parts of their range, due to habitat loss and collection for the pet trade. Fortunately, scientists now work toward breeding the animals in captivity and then releasing them back into the wild, but this process requires that we keep learning more about the snakes and their needs. After all, if scientists don't fully understand how a species gets along in the wild, then they can't ensure that the areas where they are reintroducing the snakes are suitable for them.

So, in a study led by Auburn University herpetologist Scott Goetz, the scientists presented baby eastern indigo snakes with cotton swabs dabbed with a variety of smells and then recorded the snakes' reactions.

A quick side note about snake schnozzes: Snakes have evolved a number of ways to perceive the world around them. For instance, pit vipers—such as rattlesnakes, cottonmouths, and copperheads—can sense infrared, which allows them to track warm-bodied prey. But in some species, such as the eastern indigo snake, scent detection has taken precedence, making the animals the serpent equivalent of bloodhounds.

Indigo snakes can smell in two ways. First, they can sniff the world around them using a pair of nostrils, just like you or I do when we walk past a field of flowers. But on top of that, snakes possess a supersensory organ that humans and our primate kin lost to the evolutionary ages. This is known as the vomeronasal or Jacobson's organ, and it's located in the roof of the snake's mouth.

You know how snakes have famously forked tongues? Well, when the serpents flick those forks out into the air, they are able to lick up all kinds of molecules drifting through the environment—in some cases, molecules that are heavier and easily sucked in through the nostrils. And when the snake draws its tongue back inside its mouth, it actually docks those two forks into holes on the roof of the mouth that deliver those molecules into the vomeronasal organ, like a plug sinking into an electrical socket. And snakes can perform this air-to-organ transfer extremely quickly as they sample all the scents around them.

As you might expect, snakes flick their tongues faster and more frequently at smells that interest them in some way—because it smells like something yummy, something dangerous, or someone to mate with.

Now, to gauge the snakelings' interest in various smells, the scientists came up with something known as the tongue flick attack score, or TFAS. The higher the TFAS, the more curious the snakes were about a given scent. Cotton swabs dipped in plain old water barely registered on the TFAS, for instance, while swabs brushed against feeder mice earned a score of 5. At the same time, swabs that smelled like gray rat snakes scored a 15, meaning the snakes flicked their tongues faster and more frequently at the scent of another snake than at that of a mouse. This seems to suggest indigos would prefer a meal of rat snake over a meal of mice.

Perhaps most interesting of all, swabs stinking of copperheads earned the highest TFAS score of all—a whopping 25—which means the test indigos were partial to that smell above all the others. And this was despite the fact that the indigo snakes were babies and thus had never before come into contact with copperheads or gray rat snakes.

Another experiment tested indigo snake tongue flicks when it came to a variety of pit vipers native to the indigos' habitat. And here, too, eastern indigo snake reactions to copperheads, cottonmouths, and pygmy rattlesnakes all scored highly. Interestingly, eastern indigo snakes also possess some level of resistance to the venom of the snakes they regularly do battle with. (Though, let's be careful not to say they're immune. "I guess I would think of being immune as being Superman. Like, you don't even notice it," says Steen. "And being resistant is that you're not as vulnerable to it as others might be.")

Add it all up and the eastern indigo snake is a long, strong, snake-eating machine. And there's reason to believe that it has evolved specifically to hunt, disable, and devour the spiciest of all snake species—the venomous vipers. After all, baby indigos that have never even met a pit viper seem to come with pit-viper-hunting software already installed. Their attack strategy—crush the face—seems designed to disable a venomous snake's most dangerous weapon, and the fact that they can withstand a bit of venom in the veins is a handy backup plan.

Finally, it's awfully telling that pit vipers tend to be not only sit-and-wait predators but also sit-and-wait prey. Rattlesnakes, cottonmouths, and copperheads each possess exquisite camouflage and will most often slink down and hope a predator walks past them, rather than slithering away like a racer or water snake.

Of course, the eastern indigo snake's ability to sniff out its prey all but obliterates the pit viper's cloaking mechanism—like superheroes and supervillains trading complementary blows designed to exploit each other's weaknesses.

Snakes battling snakes in a fight to the death, each with their own weapons, tricks, and evolutionary advantages—are you not entertained?!

It's OK to geek out over biology, by the way. Steen even sees it as an opportunity to turn snake hatred into snake appreciation. "Snakes are often not as dangerous as they're perceived to be," he says. "And getting into a fight with you is really one of the last things that it wants to do today or ever."

"I think that for many people, their first impulse is to run away, or for a small fraction, to try and kill the snake," says Steen. "But if you just observe these animals like you would a bird or a bobcat, you will see lots of exciting behaviors and natural history."

THE MONSTER OF MEDICAL IMPORTANCE

GILA MONSTER

TYPE: Reptile **SPECIES:** *Heloderma suspectum* **STATUS:** Near Threatened
RANGE: American Southwest through northwestern Mexico **SIZE:** Up to 22.5 inches long; weighs up to 5 pounds **LIFESPAN:** ~40 years (in captivity)

THE GILA MONSTER (pronounced HEE-la) is a foul-mouthed, sour-smelling lizard native to the southwestern United States and northwestern Mexico. Only slightly larger than a guinea pig, these hefty Halloween-colored reptiles come equipped with pebbly skin, which is made up of something called osteoderms. Think of them as bone chips sewed into skin. A living suit of armor, in other words.

Gila monsters, as well as their closest cousins, the beaded lizards, also hold the distinction of being some of the only known venomous lizards on the planet. But they don't have hypodermic needles for teeth like snakes or spiders do. Instead, their little daggers have grooves etched into the sides that can wick venom up out of their bottom jaw like a bathroom air freshener and then deliver it into the bloodstreams of their victims.

What's kind of weird is that, as consummate nest raiders, Gila monsters don't really need venom to take down most of their favorite foods. Tortoise eggs, snake eggs, and bird eggs aren't putting up a fight, after all, nor do newborn bunnies and baby ground squirrels, which (sorry to say) make up the heft of a Gila monster's diet.

All of which leads scientists to believe that the lizard's venom is mostly reserved for defensive purposes.

If you're feeling bad for baby bunnies, kindly remember that everything's gotta eat. Even monsters. And you should know that Gilas are the most infrequent of eaters, surviving on as little as three meals over the course of an entire year.

And as far as posing a danger to humans goes, don't let the idea of a venomous lizard keep you up at night. Because even the people who *want* to find these little beasts have a heck of a time running into one.

"You will probably never see a Gila monster," says Earyn McGee, a herpetologist at the Los Angeles Zoo. "Maybe if you go out into the desert and pray the night before." In all the time McGee has spent outdoors actively looking for the lizards she studies, she's only happened across a Gila monster three times. "They really are that rare," she says.

But there's also another reason people have very little to fear from these reptiles.

"Most of the time, lizards don't make noise. But these Gilas will hiss at you," says McGee. "They'll let you know when you're getting too close. So you really have to *try* to get bitten by them."

A Mouthful of Medicine

There are very, very few people on this planet who actually handle wild Gila monsters on a regular basis, and Jason Jones is one of them.

As the former state herpetologist for the Nevada Department of Wildlife, he's spent nearly a decade scrambling across the desert trying to locate Gila monsters and implant them with radio-transmitters. This allows scientists to keep tabs on the animals and reveals aspects of their life history we would have no access to otherwise.

For instance, did you know that Gila monsters spend more than 90 percent of their entire existences belowground? Or that an individual monster might roam a mile or more away from a den on those short, stubby legs? Or that multiple monsters can shack up together beneath a nice rock, especially within the breeding season?

Scientists didn't either—until they tracked the buggers over the course of years using such technologies. And yet, much with regard to this species remains an absolute mystery.

Gila monster reproductive ecology? "It's like the Dark Ages," says Jones.

Gila monster population dynamics? "It's kind of a black hole," he says.

How do Gila monster juveniles (less than one year old) survive to adulthood? "They're three to four inches long and cute as all get-out, but they're feisty," says Jones, who has only seen three hatchlings in all his time in the field. If you wanted to find 10 juvenile monsters in order to do any kind of meaningful study, he says, "it would probably take a lifetime."

Shoot, we don't even know what an average lifespan for the things might be.

"I know someone who has a Gila monster from the '70s," says Jones. At the time of this writing, that would make the animal at least 44 years old. "But we have no idea what their life expectancy is in the wild."

The one aspect of Gila monsters we actually understand rather well is how their venom affects the human body.

Remember when McGee said how downright difficult it is to get bitten by one of these creatures? Jones reinforced that by sharing that he's never once been nipped by any of the Gilas he's handled over the years—including the two monsters he used to keep as pets in his office, Sid Vicious and Nancy. What's more, in order to do his tracking studies, Jones must first intubate the animal, which involves carefully placing a tube down the reptile's throat while it's still awake so he can sedate the animal (and then provide oxygen while it's sedated).

"So you can imagine, trying to put a tube into a venomous lizard's mouth can be pretty exciting," he jokes.

The scientists also measure the animals' well-being by turning them upside down and gingerly dipping their tails into a beaker of water. This is because monsters store fat and other nutrients in their tails, so the more water a tail displaces, the healthier the animal is.

Despite it all, no bites.

In fact, Jones shared a saying often uttered by one of his colleagues in law enforcement: "The only Gila monster bites are the result of jackasses trying to grab Gila monsters." The data backs that up, by the way. In a study that analyzed calls submitted to United States Poison Centers as a result of Gila monster envenomation between 2000 and 2011, researchers found that of the 319 calls on record, only 105 calls involved humans. Many more involved pet dogs. And in 77 percent of cases where people were bitten, the Gila monster bites occurred on the victim's upper extremities. Friends, let me remind you that a Gila monster stands about as tall as an orange. And like that piece of citrus, the Gila monster is not built for jumping. So it's definitely not the case that venomous lizards are leaping out of the creosote bushes and ambushing hikers just minding their own business. In addition, fully 79 percent of bite victims were men, to the surprise of absolutely no one.

Still, given that bites do happen, we know that Gila monster venom causes an array of symptoms, including swelling, intense burning pain, vomiting, dizziness, weakness, rapid heart rate, and low blood pressure. In some cases, the lizards latch on and won't let go, which probably ensures more venom enters the bloodstream. (If you ever find a Gila monster securely fastened to one of your extremities, the experts advise you not to yank on the lizard, which will only make your wound worse, but to instead position the animal so that it has all four feet on the ground. If it still doesn't let go, try dunking it in a bucket of water.)

But there's actually another symptom that stems from Gila monster bites that people don't like to talk about—uncontrollable bowel movements. So, uh, why don't you see that listed on any of the lists of symptoms readily available online?

"I think it's embarrassing for people," says Jones. "Because I have known a few people who have at least puked, if not fully defecated themselves, within minutes [of a bite]. And then afterward they were still having problems when it comes to their bowels and digestive system. So it does *something*!"

But Gila monster venom isn't all bad. In fact, it's mostly fantastic for humans!

In the 1990s, scientists discovered that a peptide found in Gila monster venom looked very similar to a hormone found in the human gut that promotes the release of insulin and maintains glucose regulation. They were then able to create a synthetic version of that peptide and turn it into a real-life drug that helps humans manage blood sugar levels.

And we're not talking about some pie-in-the-sky hypothetical here. These drugs, known by names like Byetta, Exenatide, and Bydureon, started hitting the market around 2005. This means the chances are good you know someone who has taken medicine inspired by Gila monster venom. (Hi, Uncle Dave!)

What's more, scientists have since been able to keep building off that initial discovery, and there are now Gila monster–derived drugs that have "resulted in unprecedented improvements in obesity and diabetes management," says Lisa Beutler, an endocrinologist at Northwestern University. (Though not originally a weight-loss drug, Ozempic also hails from this line of research and was approved by the U.S. Food and Drug Administration in 2017.)

So don't let anyone tell you this lizard is lazy. Lethargic, maybe. Even though the animals peek their heads aboveground just a few times each year, their venom is out there every single day making life better for you and yours.

THE SNOT OTTER

HELLBENDER

TYPE: Amphibian **SPECIES:** *Cryptobranchus alleganiensis* **STATUS:** Vulnerable
RANGE: Eastern U.S. **SIZE:** Up to 29 inches long; weighs up to 5 pounds **LIFESPAN:** Up to 25 years

---

IN THE COLD, quick streams of the Appalachian Mountains, there lurks a salamander of unusual size. It has black, beady eyes, a flattened body, and can weigh more than three cartons of eggs. At full length, the things can grow as long as two bowling pins laid end to end, making it easily one of the largest amphibians on the planet.

Most know this organism as the hellbender, and there are currently two recognized subspecies—the eastern hellbender *(Cryptobranchus alleganiensis alleganiensis)* and the Ozark hellbender *(Cryptobranchus alleganiensis bishopi)*. The eastern variety is most widespread, with populations found from Georgia, Alabama, and Missouri all the way up to just a few hours' drive outside of New York City. The Ozark hellbender can be found—you guessed it—in the Ozark Mountains of the American Southeast.

If you're wondering where the name came from, no one is quite sure. But it seems European settlers may have been so horrified by the sight of the slimy beasts that they decided only hell could have spawned such a creature and figured the creepy monsters were hell-bent on returning there. Hellbender skin, which is usually a deep brown but can also come in a sort of leopard print, tends to be covered in lots of slime. It's thought that the ooze plays a role in protecting the creatures' soft exteriors against scratches from the rocks and gravel they live in and under. The goo may also provide some protection against infection. But one thing being snotty by nature definitely does is make them really, really difficult to grab onto.

"You can imagine how easy a meal a hellbender would be for an otter," says Caleb Hickman, supervisory wildlife biologist for and tribal member of the Eastern Band of Cherokee Indians. The amphibians have neither scales, claws, sharp teeth, nor a venomous sting or poisonous skin with which to protect themselves. But as many a freshwater scientist has learned, a hellbender coated in mucus and doing its best alligator-roll is extremely difficult to pin down. The defense mechanism has also earned hellbenders the nickname of "snot otters."

Adding to the hellbender's quirky reputation is the fact that their flanks come lined with frilly skin flaps. Unlike those of flying squirrels, these folds aren't good for gliding, but they do serve a pretty important purpose.

"These folds of skin are like an archaic version of gills," says Hickman. While lots of other amphibians get most of their air at the water's surface or absorb it from water using gills, hellbenders soak up nearly all of their oxygen through their belly rolls, which are lined with oxygen-absorbing capillaries. Hickman says the giant salamanders will even gyrate a bit so as to slosh all that cold, oxygen-rich water around and make sure the good stuff gets in all their nooks and crannies.

All of which is fine and good for the hellbenders, but to the human eye, these flaps give the animals a rather, well, odd appearance. They have also earned the amphibians yet another nickname, and this one is my favorite: old lasagna sides. Other names for hellbenders include water dogs, mud devils, grampus, and Allegheny alligators, the last of which refers to the Allegheny River of northern Pennsylvania.

You know how kids say, "Sticks and stones might break my bones, but names will never hurt me?" Well, if you're a giant amphibian with mucus oozing out of your skin, it turns out names might kind of matter.

In one study, Hickman scoured 153 years' worth of print and online media for references to hellbenders. Then, he and his co-author, Shem Unger, scored each reference on a scale that allowed them to measure sentiment, as well as how that sentiment has changed. In the end, one descriptor appeared nearly twice as much as any other word across 288 news pieces penned over the course of more than a century—*ugly*.

In fact, from 1863 to 1929, there was just one positive story written about hellbenders, while during the same time period there were 28 negative pieces and 53 pieces that scored as neutral. Now, some of this is no doubt due to the misguided notion that hellbenders are poisonous. But it also probably doesn't help that most people don't spend all day with their faces in frigid streams, like Hickman. Thus, there was once a pervasive idea that hellbenders preyed upon game fish and their eggs. Fortunately, these ideas can now be laid to rest.

"They were once persecuted for a misunderstanding that they were dangerous, and you'll still hear that sometimes by some of the mountain communities," says Hickman. "But they won't hurt you, and they don't diminish trout populations." In fact, the hellbender's favorite food appears to be crayfish.

"We know that populations have decreased in some areas. And in some areas, there are actually no hellbenders," says Hickman. Also worrying is the fact that in many of the places where the amphibians still persist, there's a strange gap in the age structure of the population. That is, scientists find full-grown adults who are reproducing and putting out lots of little larvae. But no matter where they look, juvenile hellbenders are nowhere to be seen.

The concern is obvious. What happens when the older generation passes away and there are no young'uns to replace them?

All Hail the Den Master

If you think about it, some of the most iconic and well understood wildlife behaviors have to do with courtship: bucks tangling their tines on a frosty fall morning, wild turkeys puffing up their breasts, fireflies flickering in the friscalating dusklight. But for many animals, reproduction remains intimately out of sight, and that may contribute to how we value them. After all, I think that if Tasmanian devil–size salamanders were battling each other for breeding dominance on the side of the road, dang near every one of us would put our lives on hold to watch.

The good news is that scientists now know more about the private lives of snot otters than ever before. And this knowledge may even help the general public refrain from accidentally mucking up salamander habitat—something hellbenders cannot afford.

Just like with deer, hellbender mating requires a show of force. And just like with deer, the males aren't battling over an individual lady, per se. Instead, they wrestle for a rock.

Now, you might think stones are easy to come by in a cold mountain stream. But male hellbenders require rocks with certain properties. Instead of focusing on the clarity, cut, color, or carat of a stone—like humans covet in engagement rings—male hellbenders look for large, flat rocks or boulders with little hollows underneath. Once a suitable rock is found, the male will squish himself underneath and set to work carving out a little love lair. Then it's up to him to defend that nesting space from all challengers.

Once a large male has secured his spot, he will push, bite, and generally harass any other males who try to come near. Female hellbenders are allowed in, of course, but they, too, have to prove their worth by making it past the male, who at this point is known as the den master.

Once inside the aquatic champagne room, the female hellbender lays between 200 and 400 eggs in a gooey, yellow mass made out of gelatinous strings. The male then fertilizes the baby ball, kicks out his lady of the night (hellbenders are generally nocturnal), and goes about protecting those offspring for up to 75 days. Once the paper clip–size larvae pop out of their soft shells, they will remain with the den master until the spring, at which point they will wander out into the world and mostly get mauled by fish and other predators.

The thing is, we know this is a numbers game, says Hickman. Females lay hundreds of eggs, but only a precious few will ever make it to adulthood.

In an attempt to skew those odds in the hellbenders' favor, experts at the St. Louis Zoo have been breeding and releasing both Ozark and eastern hellbenders back into their native habitats. Since the early 2000s, the program has been able to reintroduce more than 10,000 salamanders, and in 2023, all of that work was rewarded with a landmark achievement. For the first time ever, one of those captive-reared water dogs successfully reproduced on its own out in the wild.

The good news is that sentiment toward hellbenders has shifted dramatically over the years, and not just among hellbender-obsessed scientists. In fact, between

1980 and 2016, Hickman found a whopping 92 positive stories written about hellbenders, compared with just 45 neutral and three negative stories. You might even say some people are a little too into the snot otters.

"There is a pet trade for hellbenders," says Hickman. "It's a secretive group, but we know that they're out there."

Imagine that. A little over a century ago, the hellbender's biggest problem was being killed on sight because it was ugly. And today, public opinion has swung so far in the other direction, we now have to redact location details on scientific documents to prevent people from loving old lasagna sides to death.

THE JELLY-MURDERING MOUNTAIN

LEATHERBACK SEA TURTLE

TYPE: Reptile **SPECIES:** *Dermochelys coriacea* **STATUS:** Vulnerable
RANGE: Basically all oceans and coasts, except the Arctic and Antarctic **SIZE:** Up to 10 feet long; weighs up to 2,000 pounds **LIFESPAN:** ~50 years

---

"I'LL NEVER FORGET THE first time I saw one emerging from the water," says Aliki Panagopoulou, an environmental scientist with the Leatherback Trust. "You look from a distance, and there's waves, and then there's this big mass that is coming out and is moving relatively slowly."

"And then you realize, no, it's not a mountain," says Aliki. "It's a leatherback turtle."

As the largest turtles on the planet, typical adult leatherbacks can measure six feet in length and weigh around 1,000 pounds. But truly massive specimens have been seen from time to time. For instance, the largest leatherback on record measured 10 feet from its beak to the tip of its tail and weighed in at more than 2,000 pounds.

That's several hundred pounds heavier than the largest moose.

As if to match their size, these black-and-white-speckled behemoths call entire ocean basins their home. They are *sea* turtles, through and through, and while they hatch out of nests built on sandy coasts around the world, the vast majority of their lives is spent in the water. They eat, sleep, and breed in the ocean, returning to land only to lay eggs before diving back into the waves. As such, leatherbacks are the world's most widely distributed reptile.

This also means that the animals inhabit the waters surrounding virtually all of North America and have been documented nesting on Mexico's two coasts—in Oaxaca in the west and Anton Lizardo in the east.

Leatherbacks are not only the sole surviving species in their genus but also the only nonextinct creature in their entire family, the Dermochelyidae. The leatherback then is not just a mountain but, evolutionarily speaking, an entire island unto itself.

The other glaring difference between leatherbacks and most of their relatives?

"Leatherbacks do not have a shell like other turtles do," says Aliki.

Instead of a back covered in stiff plates made out of keratin, leatherbacks have tons of tiny bones that are fused together and covered in a thick layer of skin. To the touch, Aliki says it feels like the meaty part of your hand between the thumb and forefinger. The flesh can move, but it's also solid underneath. And while this all adds up to a structure that provides the turtles with some level of protection against predators, such as orcas and great white sharks, it also allows the behemoths to turn themselves into biological submarines.

Leatherbacks perform some of the most impressive dives of any air-breathing animal. The deepest on record goes to a male turtle that plunged to a maximum depth of just over three-quarters of a mile, more than a thousand feet deeper than the tallest building on Earth is tall. A different leatherback, this time a female, holds the record for the longest dive, which lasted 86 minutes. That means a leatherback could watch the movie *Toy Story* on a single breath.

But none of these feats would be possible without that huge, weird, pseudoshell.

As you go deeper and deeper into the ocean, the pressure around you grows. This is because the weight of all the water above is pushing down, silently crushing you. At the same time, any gases present in your body get squished in size (or volume), and this means air-breathing animals have a big problem known as the thoracic squeeze. In other words, the deeper you go, the more the ocean tries to crumple your chest cavity like an aluminum can.

The leatherback turtle's solution? Let it.

At a depth of around 300 feet, the leatherback's lungs and windpipe start to collapse. At the same time, the turtle's carapace—or shell—bends inward, compacting to accommodate the high-pressure world around it.

On the flip side, the leatherback's flexible arrangement allows the animal's carapace to expand when necessary. And if you're wondering why a turtle would ever need to expand, well, have you ever had to loosen your belt after a large meal?

Because for the world's largest turtle, life is just one enormous all-you-can-eat buffet.

The Last Thing a Jelly Ever Sees

Imagine if all day, every day, all you ate was watermelon.

My daughter thinks that sounds like a great idea, but she's too young to have developed a clear understanding of how an overabundance of certain foods relates to an overabundance of certain intestinal discomfort.

This is because watermelon is about 92 percent water. And the human digestive tract has not evolved to survive on watery pulp alone. But leatherbacks? Leatherbacks pig out on gelatinous organisms to exclusion. Favorites include jellyfish—which at 96 percent water are actually even more liquid than watermelon—but also other soggy, see-through creatures such as salps and pyrosomes.

So how do leatherback turtles get so colossal on a diet that's mostly nutrient-poor goo? By eating a mind-boggling amount of the stuff. By the time a leatherback reaches maturity at around 15 to 20 years of age, it will have eaten 661,000 pounds of jellies, or about the weight of a Boeing 747 airliner.

One thing that allows leatherbacks to make a living on gelatinous critters is the fact that such animals tend to come in swarms. "If there's one jellyfish, in all likelihood, there's going to be a lot more around," says Aliki.

It probably also doesn't hurt that some jellies can be surprisingly massive. The world's largest jelly is known as the lion's mane jellyfish *(Cyanea capillata),* and the suckers can measure 120 feet in length (that's a jelly longer than a blue whale).

And leatherbacks absolutely *crush* lion's mane jellyfish. One study conducted off the coast of Cape Breton Island, in northeastern Canada, found that out of more than 600 prey captures observed, between 83 and 100 percent of them consisted of lion's mane jellyfish.

Of course, many jelly species are also formidable foes, packing microscopic venom harpoons inside their tentacles. Box jellyfish are especially spicy and boast venom powerful enough to kill a human. And yet, leatherbacks soar through clouds of these creatures, gobbling them up without so much as a case of indigestion.

Leatherbacks have a number of evolutionary adaptations that allow them to live such a lifestyle. The reptiles don't have teeth, but their mouths form sharp beaks with two clefts that almost look a bit like fangs—all of which are excellent at grabbing onto and carving up squiggly prey. But it's what's inside the turtle's mouth that's truly horrifying.

Ask a leatherback to open wide and you'll find a forest of horny structures that look like a medieval torture device. In truth, these backward-facing spines are known as esophageal papillae, and each one is coated in keratin. From the sides of their cheeks to the lining of their throat and all the way down to the beginning of their stomachs, these papillae form a sort of fun-house slide of doom that holds slippery prey in place while the turtle expels all the seawater it gulped down with its meal.

The spines also macerate or chew up the jellies as they descend into the stomach. Fortunately, jellies do not have brains, and thus probably cannot process the absolute torture of being sucked down this hallway of spines before they're crammed into the stomach alongside the mashed-up remains of hundreds of their close cousins. The leatherback's belly will be so engorged by meal's end that it will actually wrap around its heart.

Given how conspicuous a reptile larger than a grizzly bear must be, it's surprising that scientists say we still know far less about leatherbacks than we do many other large and charismatic species. The animals remain difficult to breed in captivity, and younger turtles are almost never spotted out in the wild. At the same time, these gentle giants face numerous threats to their continued existence, and the International Union for Conservation of Nature lists the species as vulnerable to extinction. However, some subpopulations, such as the East Pacific leatherbacks, are listed as critically endangered.

Climate change threatens to reduce their nesting habitats, says Aliki. A warmer world may also mess up sex ratios, since leatherback eggs—which are nearly as large as tennis balls—produce females at warmer temperatures and males at cooler temperatures. Strikes from vessels, getting caught in fishing nets, and accidentally ingesting plastic bags that look like jellies are all serious dangers to the world's largest turtle.

"If you stop to think about all the threats and pressures that leatherbacks as a species are facing, you're dealing with some of the most serious environmental issues on the planet," says Aliki. And that's why she's adopted the following motto: "If you save the turtles, you save the world."

THE LETHAL MARACAS

RATTLESNAKES

TYPE: Reptile **SPECIES:** 56 species in the genera *Crotalus* and *Sistrurus* **STATUS:** Varies by species, from Least Concern to Vulnerable **RANGE:** Southern Canada through Mexico **SIZE:** Up to 8 feet long; weighs up to 15 pounds **LIFESPAN:** Up to 25 years

---

DID YOU KNOW THAT, unlike most reptiles, rattlesnakes give birth to live young?

"Oftentimes the mother will hang out with and take care of the babies for about two weeks," says Emily Taylor, a herpetologist at California Polytechnic State University.

And sometimes it's not even Mom doing the babysitting, but another adult snake instead. The snakelets are born midway through a shedding cycle, which makes their eyes foggy, so they're especially susceptible to predation and even dehydration, says Taylor. They are rattlesnakes, but they are also extremely vulnerable.

"If a predator comes, the mothers will take their body and kind of scoop them into a burrow while rattling and acting defensive around them," says Taylor.

Some reptiles can get all the water they need from the food they eat, but not rattlesnakes—they need to drink. Trouble is, lots of rattlesnake habitats have no standing water. So instead, when it rains, all the little slitherers come hither.

"They curl up and flatten their bodies and create a kind of a dish," says Taylor. "It's almost like a bowl where the water will accumulate in the fold of their skin. But their skin is also hydrophobic, or water-hating, so the water beads up, and the snakes will come around and drink a kind of circular pattern around their backs."

Sometimes, rattler mothers and babies will even sip the liquid off each other.

I'm telling you all of this for two reasons: First, it's patently adorable. But also, I think the story of rattlesnakes—of which there are 56 known species—is too often focused on bite statistics and venom symptoms and the likelihood of humans getting bitten.

That people-first framing overshadows the fact that rattlesnakes have been coiled up on the North American side of the planet for around 12 to 14 million years. And while archaeologists aren't quite sure when the first humans arrived over the land bridge that once connected modern-day Russia and Alaska, most evidence suggests it was only within tens of thousands of years from now. Which means, in the grand scheme of evolutionary time, we are little more than a passing fad to these ancient reptiles. And if you can step outside of the human-centric view, which is often guided by fear, you'd see that rattlesnakes are the scaly string that holds ecosystems together.

Rattlesnakes are what's known as mesopredators, which basically means they're the food web's middlemen. With infrared-sensing pits on their faces, these vipers can "see" the warm bodies of the critters they prey upon—even in the dark. And by removing such creatures from the ecosystem, snakes actually help keep everybody healthy.

"Rodents carry lots of diseases, like hantavirus and plague," says Taylor, naming just a few pathogens that can spill over from rodents into humans.

By the same token, one study discovered that by eating the small furry mammals that carry them, a single timber rattler could eradicate 2,500 to 4,500 ticks from an area each year. And because some tick species can transmit pathogens that cause illnesses in humans, such as Lyme disease, having rattlesnakes around may actually be a good thing.

Of course, rattlesnakes are also part of the menu for other animals. So when rattlesnakes eat rodents, they are also turning themselves into healthy, plump "snake sausages that then get carried away by hawks, mountain lions, and other top predators," says Taylor.

Coyotes, bobcats, foxes, and even roadrunners kill and eat rattlesnakes. So do raccoons, skunks, magpies, black bears, feral pigs, and other snakes, such as king snakes and indigo snakes. There's even evidence that deer, cows, horses, and bison will stomp a rattlesnake to death if startled or perhaps as a defensive maneuver.

All of which is to say, it can be a scary world out there for a relatively little reptile with no legs and only one bitey end. So let's go ahead and talk about the things that probably interest you the most: the rattles and the bites.

Rattle On

If you're ever out in the world and hear a *bzz-zzz-zzt-zzt-zzt* that sends a shiver down your spine, you should count yourself lucky. Because even if you can't see it, you've just met a snake that really, really does not want to bite you.

"Rattlesnakes actually rarely rattle," says Taylor. In fact, out of the thousands of venomous serpents she's encountered in the field, the vast majority simply freeze and hope not to be seen. Often they'll retreat to the brush or a nearby squirrel hole rather than chance a confrontation.

"It varies by species and individual personalities. Some of them never rattle. Some of them are super buzzy," she says. "But they rattle less than half the time, even when you stand next to them or pick them up or touch them."

You will want to avoid doing any of those things, of course, but Taylor is also the owner of Central Coast Snake Services, which humanely removes and relocates rattlesnakes from peoples' yards. So she has more experience than most.

Think about it from the snake's perspective. When rattlesnakes aren't predators, they're prey. And while that rattle is used as a warning of last resort, it might also help a predator zero in on the snake's location.

All of which brings us back to you standing on a trail in the middle of nowhere and hearing a sound nobody wants to hear. First, take a slow, steady step away from that nerve-racking razzmatazz. Then another, and another, until you've put at least 10 feet between you and the snake, at which point you should be safe. All the while, try to remember:

A rattlesnake can't eat you. So a rattlesnake doesn't really *want* to bite you.

Venom isn't free. Every drop has to be produced by the snake's body, which requires energy and also time. It can take a snake two weeks to stock back up after using its stores. "If they waste that venom on you, they won't have as much for the next prey that comes around," says Taylor. In fact, some rattlesnakes are downright stingy with the stuff. In around 15 percent of human bites, rattlesnakes don't inject venom at all, says Taylor. These are known as dry bites.

Now, it is not my intention to convince you venomous snakes are harmless. They are not that.

Each rattlesnake species is equipped with a unique blend of proteins molded by evolution to wreak absolute havoc on different parts of a prey's anatomy. Some snake venoms dissolve tissue or stop the heart. Others make blood clot, or prevent blood from clotting, or even assault the attacker's smooth muscles, leading to simultaneous vomiting and diarrhea. Not for nothing, but the same features that make snake venom dangerous also make them a veritable biological pharmacy for scientists who study how each protein works and then use that information to create new medicines that can save human lives.

Still, around 40 people lose their lives to snakebites in North America each year (around 35 of those take place in Mexico, with the rest occurring in the United States). But thousands more are bitten and require significant, life-saving medical attention. These cases account for around 7,000 to 8,000 people annually in the U.S., 3,800 victims in Mexico, and 100 in Canada.

And while I in no way want to make light of the pain and anguish suffered by someone who wore flip-flops in their backyard or dug in the wrong part of their garden, I do want to make the point that lots and lots of venomous snakebites in North America are our fault.

Scientists divide snakebites into two categories—legitimate, which means accidental or unavoidable, and illegitimate, meaning cases in which the victim put themselves in harm's way. (No, I'm not sure why scientists use different terminology for shark bites—provoked and unprovoked—but the effect is the same.) And let me tell you, the illegitimate ones have a bit of a type.

One study of rattlesnake bites in Arizona found that 87 percent of victims just so happened to be male. And another study—this one including a variety of venomous snake species, all of whose victims sought care at a medical center—found that over the course of 10 years, 67 percent of bites came as the result of intentional exposures. In other words, the victims included snake hunters, pet owners, or people who saw a snake in the wild and thought, "Hmm, I should probably mess with that." Also on the list, religious snake-handlers who believe their god will protect them from venom. (Note: Enough people have died while handling snakes in this way that several states have banned the practice.)

Another interesting tidbit from that study? Forty percent of the bites tracked over that decade occurred after the person had consumed alcohol. And that brings us to the kicker: According to all those U.S. hospital records, fewer than half of all venomous snake bites are what scientists would consider legitimate.

In other words, the rattlesnakes are not out to get you. Not even a little bit. After all: A rattlesnake can't eat you. So a rattlesnake doesn't really *want* to bite you.

THE TOAD TRIP

SONORAN DESERT TOAD

TYPE: Amphibian **SPECIES:** *Incilius alvarius* **STATUS:** Least Concern
RANGE: American Southwest and northwestern Mexico **SIZE:** Up to 7 inches long; weighs up to 10.6 ounces **LIFESPAN:** 10 to 20 years

---

DO NOT LICK the Sonoran Desert toads.

Also, do not eat the toads, no matter how appealing their glistening, forest-green skin looks. Nor should you rough-handle the toads, which at around seven inches in length are one of the largest toad species in North America. And while we're at it, it would be great if you could refrain from relocating the toads, running the toads over with your car, or milking said toads' poison glands and then inhaling their defensive, psychoactive compounds in the pursuit of enlightenment, medical treatment, or a good time.

Of course, most of us would never dream of licking a toad. But when you read a list that specific, well, you know they've all been done before.

Sometimes called the Colorado River toad, the Sonoran Desert toad is the only animal on Earth known to produce easily accessible concentrations of a powerful psychedelic compound known as 5-MeO-DMT—also known as the God molecule.

The molecule gets its name because people who have purposely taken the toad toxin describe psychological experiences like riding a rocket to the center of the universe, the disintegration of their egos, or the transcendence of time and space.

But toad-lickers should be warned—other side effects include existential dread, horrific flashbacks lasting for weeks, and, in a few extreme cases, death.

That's right, the Sonoran Desert toad can kill. Which shouldn't be that surprising. That's why toad toxins evolved, after all.

"A lot of dogs and cats die every year from biting and licking these animals," says Robert Villa, president of the Tucson Herpetological Society. "All toads produce cardiac glycosides from their skin. And those are chemicals that relax your heart muscles."

The good news is that if you do not go in search of a life-threatening experience, then you are very, very unlikely to be harmed by a Sonoran Desert toad.

This is because toads are poisonous, not venomous, says Villa, who is also the community outreach assistant at the Desert Laboratory at the University of Arizona. And while the difference between what's poisonous and what's venomous may seem like semantics, it's actually rather important for understanding the risks imposed by various forms of wildlife. So let's break it down.

One easy way to think about it is that both poison and venom are toxins, says Villa. The difference is in how each toxin is delivered.

For poisonous animals, such as the Sonoran Desert toad, those toxins weep out of glands on the surface of the amphibian's skin. And the poison is only really dangerous when a predator tries to eat that animal. In those cases, the liquids come into contact with the mucus membranes inside the predator's mouth, or—if an animal is digested—introduced into the bloodstream.

Long story short, unless you try to gobble a poisonous animal or accidentally rub its secretions into your eyes, nose, or mouth, you should be good as gravy.

On the flip side, venomous animals have evolved fangs, spines, and stingers that allow them to send their toxins out into the world, like liquid emissaries of pain. In most cases, venomous animals don't actually want to attack humans, but they still pose a greater threat than poisonous animals, simply because they can bring the fight to us.

That said, there are exceptions to every rule. For instance, when the Spanish ribbed newt is in danger, it flexes its chest really hard and fast, which causes some of its ribs to break through its skin, where they can wick up poison and transport it to a predator's soft, fleshy mouthparts. And then there's the Asian tiger snake, which scientists believe is able to steal the toxins found in the poisonous toads it dines upon and then secrete them from special glands near its head. The snake also has fangs and venom glands in its mouth, which means both it and the newt qualify as poisonous *and* venomous. It's yet another example of how people like to put things in tidy little bins, but nature dumps those bins out and mixes everything up when we're not looking.

A Miraculous, Magical Amphibian

Frogs, toads, newts, and wormlike animals known as caecilians are all amphibians, or animals that spend part of their lives in water and part on land. And while lifestyles differ across species, the name for these animals comes from the ancient Greek word *amphibios,* which means "both kinds of life."

"Amphibians sit between fish and reptiles, insomuch as their water needs are pretty important," says Villa. As such, it's honestly pretty strange that amphibians can exist at all in a place like the Sonoran Desert, where temperatures can sometimes seethe into the range of 118°F.

"To be a desert amphibian means that you have to sort of break the rules," says Villa.

This is why adult Sonoran Desert toads bury themselves underground from October to April. Similar to hibernation, this strategy is known as aestivation, and it allows some animals to wait out long periods where conditions are hostile or food is scarce. During aestivation, toads not only stop eating, but their metabolisms also slow and they consume much less oxygen than normal.

After months spent frozen in time, the toads begin to wake at the onset of the summer rainy season when the Sonoran Desert is set to explode with life. Creeks that have been bone-dry for months begin to fill with water, wildflowers blossom, and amphibians and insects crackle back into existence, drawing in predators of all shapes and sizes who will feast upon them.

"The monsoon, for us, it's like second Christmas," says Villa. "We look forward to it every year, and it's crucial for the Sonoran Desert."

For the Sonoran Desert toads, it's a race to make up for lost time. As the adults emerge from the ground like mud-caked zombies, the night air becomes saturated with the amphibians' calls, which sound eerily like ferryboat whistles. Males develop calluses on their knobby little thumbs, the better to hold onto females as they mate. And females deposit clouds of gelatinous eggs into puddles and ponds. After hatching, pollywogs (or tadpoles) develop for about a month before undergoing a metamorphosis in which they will say goodbye to their tails and hello to a set of powerful leaping legs.

And despite having a defense mechanism that can stop your heart or cleave your psyche in two, Sonoran Desert toads don't mind being around people.

"I would say they're slightly more common in areas near people, because people subsidize their existence with water," says Villa, referring to human-made ponds for livestock and agriculture, as well as suburban water features in yards, pools, and the like. "And the electrical lights that people have attract insects that the toads like to eat." Remember, the animals are harmless if you're careful—if you need to move one out of the driveway or away from a family pet, simply wash your hands thoroughly before and after handling.

As the largest toads native to North America, Sonoran Desert toads, says Villa, can feast upon all kinds of creatures, from beetles, tarantulas, wasps, and centipedes to small vertebrates. Some have even been known to ingest used ammunition casings, so indiscriminate are their tastes.

Unfortunately, Villa and other experts believe Sonoran Desert toad populations may be on the verge of collapse. The reason? In both Mexico and the U.S., toads are being illegally removed from the wild each year so malicious actors can extract the animals' so-called God molecules (even though these molecules can be easily replicated in a lab, Villa points out).

While some people have tried to justify their toad harvesting by claiming various Indigenous peoples of the Sonoran Desert historically used toad poison for health or ritualistic purposes, Villa says, "There's just no evidence whatsoever."

What we do know is that the Yoeme (yo-EH-me) or Yaqui (YAH-kee) people of southern Sonora viewed the toad as a bringer of rain and would even adorn the amphibians with symbolic ribbons and parade them around town in an attempt to ensure a good growing season.

While I don't want to totally dismiss the emerging promise of psychedelics in modern medicine—the psilocybin found in some species of mushrooms, as well as other naturally based psychedelics, has captured the interest of leading medical facilities, such as the Johns Hopkins Center for Psychedelic and Consciousness Research—such practices are best left to the professionals. Because even if the threat of an excruciating and terrifying death by toad poison doesn't dissuade you, maybe this will: In this region, some people believe the Sonoran Desert toads, as symbols of rain and fertility, have the power to curse individuals who harm them. "Toads are magical creatures," says Villa.

Toads spent millions of years evolving toxins to keep other animals away, and yet now it's those same toxins that draw (some of) us near. Do yourself a favor, though.

Do not lick the Sonoran Desert toads.

Sockeye salmon
(Oncorhynchus nerka)

PART IV

FASCINATING FISH

THE ANCIENT, ARMORED AIR BREATHER

ALLIGATOR GAR

SPECIES: *Atractosteus spatula* **STATUS:** Least Concern **RANGE:** American Midwest, Southeast, and South through Veracruz, Mexico **SIZE:** Can exceed 8 feet 5 inches long; can weigh more than 300 pounds **LIFESPAN:** More than 100 years

---

THE ALLIGATOR GAR is a long-lived apex predator with a snaggletoothed grin and a body that won't quit. Yet this species is criminally underknown. So let's bring you up to speed.

Gar (rhymes with *car*) are easy to identify because they all have torpedo-shaped bodies and long snouts. While usually seen in freshwater streams, rivers, and lakes, gar can also survive in brackish, or semisalty, coastal waters, as well as the ocean. Today, there are seven known species of gar found from southern Canada all the way to Central America and Cuba. Of them all, the alligator gar is by far the largest.

Just think about how big these honkers get. At around eight feet in length, the largest alligator gar would be longer than any basketball players ever to play in the NBA. And at more than 300 pounds, these underappreciated fish would outweigh the heaviest mountain lion ever recorded. Put those stats together, and you can call the alligator gar the second largest freshwater fish species in North America. Only the white sturgeon surpasses its gigantitude.

Now, how does a fish get to be named after a giant reptile? That part is pretty simple.

"It literally looks like if you took an alligator and replaced its legs with fins," says Solomon David, a gar expert at the University of Minnesota.

Much of that resemblance is probably due to the alligator gar's grin. Inside the mouth, these fish possess not one but two rows of sharp, needlelike fangs. The inner row of teeth can actually grow longer than the outer and produces the teeth-everywhere-all-at-once kind of smile only a mother could love. But this mangled mouth has a purpose!

Alligator gar are sit-and-wait predators capable of launching themselves forward and grabbing hold of slippery prey. They mainly eat fish, but once they get to several feet in length, they have been known to snatch everything from frogs, snakes, and turtles to small mammals and even birds. Every once in a while, a fisher is surprised to start gutting an alligator gar on a dock only to find a bellyful of ducklings.

"They are opportunistic carnivores," says David. "They are primarily going to be feeding on fish, but if something else presents itself, well, they'll go for it."

Don't worry. Humans are one of the few things not on the menu. That's because a gar can't tear off chunks of flesh like an alligator or a great white shark can. If gar can't swallow a meal whole, they don't bother.

Apart from growing as long as a sectional couch, what's really fascinating about this mighty fish family is that it's been doing its thing for about 157 million years. And while the individual species have diverged and changed a bit, the very basics of gar-ness remain as they were way before *Tyrannosaurus rex* loped across North America.

Think about that for a second. Not only have gar been on this landmass since the time of the dinosaurs, but they survived the 7.5-mile-wide asteroid that crash-landed in the Yucatan Peninsula. After the asteroid hit, an infrared pulse rocketed across the globe, raising air temperatures about 500 degrees Fahrenheit, says Riley Black, a paleontologist and author of *The Last Days of the Dinosaurs*. That means much of the planet would have been on fire after the cataclysm. And the alligator gar's ancestors would have been right in the thick of the unfolding apocalypse.

"If you go to the Hell Creek Formation, which would have been the stomping ground of tyrannosaurs and triceratops, we find a ton of gar scales," says Black. "We know they were doing really well."

One of the reasons gar scales have been able to survive the ages is because they're not like the shimmery, interlocking bits you've seen fall off goldfish or bluegills. In fact, gar scales are more like medieval armor. Each one is covered in ganoine, a multilayered, mineralized tissue kind of like tooth enamel, which is the hardest substance our bodies produce, says David. This is why gar scales are sometimes called ganoid scales. It's also why if you were ever trying to cut into one of these animals, you'd want to put the knife down and pick up a hacksaw.

Now, while a suit of armor is helpful for warding off predators and maybe even some space-rock-induced climate change, it's not quite enough to get a fish through the end times. Fortunately, gar sport a few more lifesaving tricks. Remember how gar can live in fresh or saltwater? That's because they are kind of wishy-washy about how they breathe.

"Gar are what we would call facultative air breathers," says David.

In other words, they can breathe underwater using their gills, but they can also gulp air from the water's surface, and then extract oxygen out of that air using their swim bladder. This allows the animals to survive in all kinds of different environments, from oxygen-rich, cold-water ecosystems like swiftly moving rivers and streams, to hot, stagnant swamps and pools. Saltwater, freshwater, and anything in between—if it's wet, it's fit for a gar.

Another boon in a disaster? Gar are ectotherms, or what we sometimes call cold-blooded animals. And that means they can dial their metabolisms way down in times of famine. Of course, this became more than a neat party trick during the event that annihilated 75 percent of all living things on this planet.

"We often look at these kinds of animals as somehow lesser, or not as interesting, or less behaviorally complex," says Black.

"But the fact is, if you're an ectotherm, and you don't have to eat a whole lot, and your body temperature is regulated by the environment around you, those organisms can persist for a really long time," says Black. "They can handle a lot of shake-ups."

Unfortunately, after shrugging off an asteroid, there's one thing that could actually wipe the hardy alligator gar off the planet for good.

Us.

Gar-gantuan Threats

Despite the fact that full-grown gar have no real predators, these fish are still very much in danger across much of their range. And there are a few reasons for this.

For starters, people tend to drain swamps and wetlands to make way for shopping malls and housing plans. Which are things we need, sure. But fish need homes, too.

Second, humans do this thing where we set up dams, sometimes to control flooding or to direct its flow but also to generate electricity, which again, is a thing we need. But imagine being a fish that's swum up the same river to reproduce for the past 75 years and then, all of a sudden, one day there's a giant wall of concrete standing between you and the future of your species.

"Freshwater systems are some of our most valuable natural resources," says David. "But as we dam rivers and drain floodplains, we're destroying those habitats that gar and many other species need to survive. These are breeding areas, nursery habitats, and also feeding grounds."

Finally, gar are sometimes killed on purpose by fishers. Some of this is due to their size and tenacity when they're caught on a line—trophy fishers and bow fishers prize the largest gar specimens because they put up a fight and even lurch out of the water, like swordfish. But what's perhaps even sadder is that lots of folk have been raised to believe that gar are worthless "trash fish," or fish not good for eating.

First of all, that's silly, says David, because he's eaten gar lots of times, and they're delicious (just don't eat the eggs, he warns, because gar caviar is poisonous to mammals).

But second, alligator gar and their cousins are just as much a part of freshwater ecosystems in the American Southeast as any other native fish. In fact, as top predators, they're indispensable actors in the greatest show on Earth, otherwise known as the food web. And as you now know, gar had been performing those top-dog duties for millions of years before people showed up.

Finally, David says that the stigma against gar is so strong in some places, that people have actually been known to catch gar by the dozens and toss them onto the riverbank to die. Some even believe killing gar is good for the ecosystem, because it saves all the smaller fish the predators would eat.

In the 1930s, the Texas Game, Fish, and Oyster Commission even rigged up a boat that could send an electrical current through the water, zapping everything below. They called it the Gar Destroyer.

If you ask me, it seems like a pretty silly way to treat a living legend. Alligator gar get bigger than us, live longer than us, and have a claim to this continent older than the sands of Texas, which lay beneath an ocean when these breathtaking behemoths first swam onto the scene. If you can't appreciate them for the superpowered survivors that they are, then at least leave them well enough alone.

THE CARNIVOROUS CHAMELEON

GREAT WHITE SHARK

SPECIES: *Carcharodon carcharias* **STATUS:** Vulnerable **RANGE:** Basically all North American coasts, temperate and subtropical waters worldwide **SIZE:** Up to 21 feet long; weighs up to 5,500 pounds **LIFESPAN:** Up to 70 years

---

THE GREAT WHITE SHARK is the largest predatory fish on this planet. One of the largest individuals ever recorded is a female known as Deep Blue who might weigh as much as 5,500 pounds, which would be like if you took half a dozen grizzlies and stuffed them into an armored, hydrodynamic torpedo that can slice through the sea at 35 miles an hour. Full-grown great white sharks can reach lengths of more than 20 feet, longer than a giraffe is tall.

While you might associate great whites with places such as Australia or South Africa, these predators are equally at home on North American shores. Great whites roam from Canada's Newfoundland to the Gulf of Mexico and the Caribbean. And in the Pacific, you can find whites from Alaska all the way down to California's southern coast, where San Diego has become a hot spot for great white shark juveniles. In fact, Mexico's Guadalupe Island, just off the Baja Peninsula, used to be one of the best places on Earth to see great whites in all their glory until it was closed to tourism.

Named for their unmistakable underbellies, great whites sport gray or brownish coloration on their topsides. Look closely at where the two colors meet on the shark's body and it almost looks like a watercolor painting, and each shark's swirls of color convergence are unique. Overall, this combination of light and dark—called countershading—allows the giants to somehow disappear into their surroundings when viewed from both above and below.

Even more amazingly, scientists have started to collect evidence that these iconic colors are not set in stone. In the lab, special color-containing cells called melanocytes extracted from living great whites were shown to expand and contract when doused with adrenaline and other hormones. This may explain why individual animals sometimes appear lighter or darker in complexion.

It's quite a feat for a predator of this size to hide itself in the open ocean, but somehow it works, because whites prefer to ambush their prey from below. When prey is lingering at the surface, great whites can strike so fast and so hard that their entire bodies rocket out of the water and into the air—a wildlife spectacle known as breaching.

When they're young, whites survive on fish, rays, lobsters, and crabs, but as adults, whites eat seals, sea lions, dolphins, rays, marine turtles, and even other sharks.

As popular as great white sharks have become in recent years—what other animal receives weeks of dedicated television programming on multiple channels each summer?—there is still much we don't know about these gigantic and uber-charismatic fish. For instance, while we first captured video of a newborn great white shark in 2023 off the coast of California, no one has ever seen a great white actually being born—despite the fact that great white sharks give birth to live young that measure about five feet in length! Likewise, though fishers have provided a handful of accounts of great whites breeding, no scientist has ever documented these sharks in the act of copulation.

Sharks—They're Just Not That Into You

There's a lot we don't know about this magnificent animal, but much of what we do know, we misinterpret or misunderstand. The first and most critical mistake we make about great whites is thinking that they want anything to do with us.

"Seals, and sea lions, and other marine mammals are way more nutritious than we are," says Melissa Cristina Márquez, a shark scientist and author of *Mother of Sharks*. "I mean, they've got the blubber that sharks actually need, whereas for us, we're not nutritionally beneficial to them whatsoever.

"We're like the oatmeal raisin cookies of the ocean that nobody wants."

Okay, first—ouch. And second, why?

Look across the animal kingdom, and you will find plenty of examples of predators adapting to new food sources. After eating all the deer on an island in Alaska, wolves there began to hunt otters. In Lake Erie, water snakes were heading toward extinction, until they discovered they could eat a newly arrived and invasive fish from Asia known as the round goby. (Now the water snakes are thriving and have actually been removed from the endangered species list!) And of course tons of wild animals, including bears, raccoons, crows, and ants, have had no problem adapting to living near humans and eating all the handouts we provide.

Furthermore, millions of people go to the beach every day, and many of them surf, scuba-dive, snorkel, Jet Ski, or otherwise offer themselves up on a proverbial platter. So why don't great whites regularly take advantage of all the opportunities we give them to eat us? Especially considering how relatively helpless humans are in the water?

To understand, Márquez says you need to think about things from the evolutionary perspective.

Sharks are super old, as species go. Whereas humans came onto the scene around 130,000 years ago, sharks have been around for something like 450 million years. This means sharks predate the dinosaurs. Sharks predate *trees*. Heck, sharks are older than Saturn's rings!

And for all those hundreds of millions of years that sharks have spent honing their killing craft in the world's oceans, humans simply weren't there. And even though we're there now, dressed in our finest teeny-weeny bikinis, these predators simply haven't evolved to pay us any mind.

Think about it. There are nearly 8 billion humans on this planet, and yet in 2022, there were 57 unprovoked shark attacks, and only five were fatal. This means the chances of being killed or even bitten by a shark are astronomically, shockingly low.

"We're just not enough of the fabric of the ocean to be considered a meal item," says Márquez.

On the flip side, humans kill an estimated 100 million sharks every year, usually as a result of fishing. We often kill sharks as a bycatch, meaning we kill them accidentally while fishing for something else. Today, fully one-third of all shark species known to science are threatened with extinction.

By all accounts, it's sharks that should be afraid of humans. But it's another mammal that haunts the great white shark's nightmares. And that's the killer whale.

At first blush, orcas may seem like the more cuddly of the two species—perhaps because their black-and-white coloration reminds us of penguins and pandas, or maybe it's because orcas always look like they're smiling. But don't be fooled. All evidence points to orcas being serial killers—of sharks, at least.

For starters, orcas can be as much as twelve feet longer than the absolute largest great white. What's more, they can outweigh the sharks by nearly 15,000 pounds.

And there are often more than one of them. Orcas famously hunt in packs, which is why some call them the wolves of the sea. And great white sharks, for the most part, are loners (I say "for the most part" because tracking tags are starting to reveal that some great white sharks are more social than others, and this area of research is gaining momentum). In most cases, though, a solo white doesn't stand much of a chance against a team of black-and-white behemoths.

But this is more than a hypothetical battle of the beasts. Recent evidence shows that when orcas swim into town, great white sharks either flee or, later, wash up on shore with holes in their side and their livers missing. Yes, despite not having hands or surgical tools, orcas have apparently figured out how to squeeze the most dense, nutrient-rich piece of a shark out of its body like toothpaste out of the tube.

Researchers have also found evidence that the sharks can hear or perhaps smell the whales coming, and when they do, they abandon the area for at least a month and in some cases up to a year, so strong is their respect for the orcas—and, presumably, their terror.

South Africa used to be known as one of the best places in the world to see great white sharks. But in recent years, the predators have all but vanished. Why? Scientists say there are a number of possibilities—including reduced availability of prey—but some blame orcas for the disappearing act. And indeed, in 2015, a pair of orcas nicknamed Starboard and Port (due to the way one of the whales' dorsal fins bends to the right and the other bends left) moved into the area and started murdering great whites. In recent years, the duo have even started removing livers from other shark species now that the great whites have moved eastward. In 2023, Starboard and Port are thought to have killed 17 seven-gilled sharks in a single day.

Now, I'd hate to leave you thinking great white sharks are pushovers, because even with the recent targetings by killer whales, whites remain among the most formidable meat eaters on the planet. In fact, scientists now believe that great white sharks influenced the evolution of arguably the most dominant ocean beast to ever live—the giant prehistoric shark named *Otodus megalodon*.

A study published in 2025 counted growth rings on the megalodon's vertebrae and determined that it grew extra quickly in its first seven years of life—right up until the point when it would have been big enough to be safe from great white sharks. Furthermore, the "Meg" disappears from the fossil record relatively quickly after the great white rises to prominence, possibly because the smaller, faster whites were able to outcompete their legendary predecessor.

Game recognizes game.

THE CHILD OF PANGAEA

PACIFIC LAMPREY

SPECIES: *Entosphenus tridentatus* **STATUS:** Least Concern **RANGE:** Western coast from Alaska to Mexico; also in waters off Asia **SIZE:** Up to 33 inches long; weighs up to 1 pound **LIFESPAN:** Up to 15 years

---

BEFORE YOU JUDGE the Pacific lamprey by its blood-sucking, fang-filled face and parasitic lifestyle, its gray, corpselike skin, or snaky-eely wriggling, I want you to consider that every one of these characterizations is completely arbitrary. And that with even a slight shift of perspective, every creature on this planet could look good or bad. Beautiful or ugly. Important, dangerous, disposable, or indispensable.

I mean, yeah, when a lamprey latches onto a piece of glass and you can see all the yellow chompers that line its suction-disc, it might summon a bit of fear or disgust. But let's be fair here—how many of you would like to be judged by the charisma of your inner mouth?

Frankly, the Pacific lamprey is deserving of a lot more respect than it's been given in certain quarters. These unusual fish have neither scales to protect themselves, nor jaws with which to chew their food, nor bones to hold their wet-noodle bodies together.

And do you want to know why the Pacific lamprey lacks all of these seemingly basic, fishy things? It's because lampreys hail from a family of fishes that broke off from the rest of the animal kingdom so long ago that things like fish scales and bones didn't even exist on this planet yet.

When lampreys were coming onto the scene 450 million years ago, all of the land on Earth was smudged together into a giant supercontinent known as Pangaea. The lamprey lineage squirmed into existence before berries, seeds, and flowers. It predates the creation of the oil that makes the gasoline that fuels your car. These fish swam from mountain peak to ocean deep before, during, and after the reign of dinosaurs. In fact, by the time an asteroid appeared out of space and nearly wiped out life on this planet, the lamprey family had already survived three other mass extinctions.

These days, a lot of the Pacific lamprey's bad press comes from the way its Atlantic cousin, the sea lamprey *(Petromyzon marinus),* invaded the Great Lakes and collapsed trout fisheries. I mentioned earlier that many lamprey species are parasites, and it's true. As adults, lampreys clasp onto larger fish with their suction-cup-like mouths and then use their cheese-grater tongues to grind up those fishies' flesh so they can suck their bodily juices. And all of that either kills the host fish outright or weakens it to such an extent that it's no longer fitting for humans to catch.

But look, even this isn't really the lamprey's fault! After all, sea lampreys couldn't get into the Great Lakes until 1829 when humans dug the Welland Canal as a way to move ships around that giant, impassable riverine barrier known as Niagara Falls. By the way, the Great Lakes are just 32,000 years old—tops. Which means they're little more than a new fast-food chain to these primeval fish. And when we installed the equivalent of a drive-through window, the lampreys just did what lampreys do—swim up freshwater streams to feed and breed.

If you're thinking that sounds a lot like salmon, you're right. Lampreys are anadromous, which means they spend part of their life cycle in freshwater and part in saltwater. Unlike salmon, however, lampreys don't have pectoral fins on their sides with which to power their migration. And this means that the lampreys must travel hundreds of miles inland using only their slithery swimming motion or their suckers, which allow the ancient fish to scale waterfalls and also hitch rides upstream on other swimming things. In this improbable way, some populations of Pacific lampreys shuffle, slither, and sucker-hike from the coasts of Washington and Oregon all the way up to the Columbia River's headwaters in the Rocky Mountains of Idaho!

At least, they used to. Humans have spent the last century and a half building dams all over the Pacific lamprey's chomping grounds (the species ranges inland from California to Alaska and out at sea all the way up through the Bering Strait and even as far as Japan). Which means a Pacific lamprey must now scale as many as eight gigantic concrete dams to get where it needs to go.

Fortunately, some friends have taken it upon themselves to give the lampreys a lift.

Guardians of the Lamprey

Stop by the Willamette Falls Lamprey Celebration in Oregon in early July, and you'll find Elaine Harvey with a sharp knife in her hand and blood on her cutting block. Behind Harvey, filleted lampreys hang on wooden skewer racks, almost like wind chimes. This helps the meat dry quicker and more evenly so that it can be stored over the winter.

"It's an art really, and I'm not even that good at it," she says humbly. "But I have my aunts who come in. They're the professionals."

As the Watershed Department Manager for the Columbia River Inter-Tribal Fish Commission, it's Harvey's job to look across the region and identify threats to all the living things that depend on freshwater, including, and especially, Pacific lampreys.

"My family are fishermen," says Harvey. "Thirty years ago, I would go fishing with my mom and dad and we would catch lots and lots and lots of lampreys. And now we go there, and you can't ... They're just not there."

So, what happened? According to Harvey, the Pacific lamprey's downfall mirrors declines in water quality, water quantity, and an increase in dams. And this is all a rather big problem, because Harvey and her people depend on Pacific lampreys.

Harvey is a citizen of the Yakama Nation, from the Kamiltpah and Wilawitis bands. And to her people, this fish, also known as *asúm* in the Sahaptin language, is considered sacred. In order to be able to breathe while also latched onto another animal, Pacific lampreys have evolved a different sort of gill than most other fish, one that is closed off from the mouth and instead opens through the seven holes on the side of their body. And in the Wáashat religion, which is shared by the Yakama as well as the Umatilla, Nez Perce, and Warm Springs peoples, the number seven is representative of the seventh day of the week, which is also considered to be holy.

The other reason Pacific lampreys are held in high esteem is that the adult fish make their yearly migration runs after the salmon. Historically, the Yakama relied on the lamprey run as a vital, life-sustaining resource. Pacific lampreys are oily, nutrient-dense fish. When dried, they're also prized teething toys for babies, says Harvey, as well as one of her culture's traditional "first foods," along with staples such as salmon, venison, buffalo, huckleberries, and bitterroot. After reaching a certain point in their lives, or encountering illness such as cancer, some Yakama elders only eat first foods, she says.

Of course, humans are also not the only animals who benefit from lampreys. "What we say and believe is that the fish is bringing all those ocean nutrients back from the ocean to the interior, to our watersheds," says Harvey. As with salmon, lampreys die after they reproduce, and their decaying flesh feeds innumerable macroinvertebrates, which in turn feed young salmon, who feed predators, such as otters and eagles. "Everything works in cycles," says Harvey. "Lamprey have a purpose in our ecosystem, in our life, in our culture. But when you take that away, it affects other parts of the system."

And that's why, in 2007, a team of scientists from the tribes, as well as state and federal agencies, started catching adult Pacific lampreys in the main stem of the Columbia River and transporting them to various sites along the upper reaches of the watershed—places where Pacific lampreys had either completely disappeared or were hanging on by a thread. And after more than 10 years of this time-consuming, painstaking work, the results have started to flow in, like silver shimmers boiling at the base of a waterfall.

For starters, the tribes' work has allowed for greater scientific understanding of the species than has ever existed before. We now know that Pacific lampreys spend an average of 6.7 years as filter-feeding juveniles, buried in freshwater sediments like worms in a garden, before heading out to sea for the first time. Once there, the animals spend an average of five years in the ocean before returning to land.

What's more, the translocations have been wildly successful. Not only did the transported lampreys reproduce, but their spawn survived long enough to make it out to the ocean. In 2021, the Columbia River Inter-Tribal Fish Commission even confirmed that the descendants of original transplants had returned to the Bonneville Dam as adults. And they did so at numbers exceeding what the most hopeful of models predicted.

Clearly, the Pacific lamprey species and its migration are not out of danger yet, and care is being taken to make sure that annual harvests don't exceed what the population can replace. But 2023 offered another glimmer of hope—the number of lampreys passing through the Bonneville Dam exceeded the 10-year average by nearly 20,000 animals.

As Harvey says, "Everything works in cycles."

THE SHAPE-SHIFTERS

SALMON

SPECIES: 5 species in subfamily Salmoninae **STATUS:** Near Threatened **RANGE:** West Coast from California to Alaska, East Coast from Connecticut to Greenland **SIZE:** Up to 58 inches long; weighs up to 120 pounds **LIFESPAN:** Up to 7 years

---

SOME OF THE CRITTERS in this book may be unfamiliar, but most of us already know what salmon are: red-fleshed fish that we most often encounter as blackened salmon entrées, salmon sushi, smoked salmon, and even salmon burgers.

But delicious though these fishies may be, reducing them to little more than a menu item is a great injustice. Because out at sea or in rivers, salmon are the magnificent middlemen that connect vast and varied ecosystems.

Take the Chinook salmon, the largest of the six species of salmon native to North America. Also known as king salmon due to their colossal size, once upon a time, this species would grow nearly six feet long and weigh 126 pounds. That's a fish bigger than the average seventh grader.

Or it would have been. These days, a more typical Chinook would measure around three feet in length and weigh 30 pounds. The reasons for the decline are many, but include pollution, deforestation, diseases, and parasites introduced by fish farming, overfishing out at sea, and dams that block the animals' annual migration.

In addition to Chinook, other salmon species include the sockeye, coho, pink, and chum salmon. In the East, there's also the Atlantic salmon, which is the species you're most likely to have bought if the fish on your plate was farmed.

The salmon life cycle is perhaps one of the strangest in nature. In most species, salmon begin life as eggs nestled into the gravel of a cold mountain stream. After hatching, these tiny fish are known as alevins, and they aren't actually very fishlike. In fact, they're still attached to the yolk from their eggs, which acts like a fanny pack of snacks that keep the little ones nourished as they hide out at the bottom of their streams.

Basically everything can kill the alevin at this point, including too much light and even vigorous swimming, because if that yolk sac gets damaged, the little ones are too weak to feed on their own.

After the external yolk is depleted, the small fries swim up and gulp some air. (By the way, they are actually called fry at this state. That's where the phrase "small fry" comes from!) Without the air, the fish are heavier than the water around them, which helps them hide in the muck, but now they need to be able to swim around and search for food, and the extra gas in their swim bladders allows for better buoyancy control.

Over time, the fry shed their dusky browns and greens for a layer of shimmery silver scales. They are now known as smolt, and that means it's time to bolt downstream and, eventually, into the bays and estuaries where rivers meet the ocean. Here, they hide, gorge, grow, and undergo a metamorphosis that's every bit as bonkers as that of a tadpole turning into a frog or a caterpillar into a butterfly.

"The fish have to undergo tremendous osmo-regulatory transformation before they can go from being freshwater fish to saltwater fish," says Andrea Reid, an Indigenous fisheries scientist and conservation biologist at the University of British Columbia.

Without getting too deep into the physiology, just imagine what would happen if you took almost any other kind of fish from a freshwater ecosystem and dropped it into the ocean. "Those fish would die in a hurry," says Reid, who is also the leader of the Centre for Indigenous Fisheries and a citizen of the Nisga'a First Nation of British Columbia's Nass River Valley.

But changing your ion-processing channels to be able to breathe in saltwater takes time, which is why smolts chill in those brackish between-waters for a while as their bodies adjust. Next, the shimmery salmon head out to sea, where they fill their bellies with other fish, eels, plankton, krill, and whatever else they can chase down as the fish—which were born in the mountains—now roam up to 1,000 miles out into the wild blue ocean.

Then, after a few years out at sea, these masters of transformation just turn around and go right back where they came from (fish that swap freshwater for salt and back again are known as anadromous fish). Of course, this also means rewiring their biology all over again to be able to breathe freshwater—a feat frogs, butterflies, and other metamorphs can't claim.

What's more, the fish also shape-shift their appearance in the process. For instance, sockeye salmon shine like silver coins as seagoing adults, but when they get the urge to breed and return to freshwater streams, their heads turn forest green and their bodies flush scarlet. Males also sprout an assortment of tiny, jagged teeth on jaws that elongate and bend inward, like crab claws. Their stomachs shrivel up as they stop eating altogether, and fat melts away as they fight against the current and hazards aplenty.

"People refer to it as 'the gauntlet,' because not only do they have to avoid fisheries, but they have to evade predators; they have to climb waterfalls and work their way up rapids," says Reid.

And remember, all of this is playing out over the course of months and many, many miles. For example, the longest known salmon migration occurs on the Yukon River, which begins in the delta on the western side of Alaska and ends in Teslin, Canada—a riverine distance of nearly 2,000 miles. But how do the salmon know where to go?

"It really comes down to this bouquet of scent," says Reid. "They smell their way home."

And not without great personal cost.

"Sometimes you arrive on spawning grounds, and you'll see salmon that are missing part of their caudal fin. Or sometimes they have this gaping wound on their head, and you can almost see their brain," says Reid. And yet, they continue to wriggle back to their place of origin. "They are so *fierce,*" says Reid.

Eventually, the fish arrive at their spawning grounds, literally falling apart, where the females carve little divots in the gravel and lay their eggs, which the males quickly and thoroughly fertilize. With nothing left to do, the salmon die (dying after reproducing once is known as semelparity, and it's a trait shared by lots of animals, including fireflies and praying mantises).

But death is just a different kind of beginning for the marvelous metamorph fish.

A Salmon's Second Life

When you or I die, our bodies will most likely either end up pumped full of chemicals and buried in the ground or burned to ash and placed on a mantel or cast out into the ocean. The point is, our mortal flesh will be mostly wasted, as far as nature is concerned.

But salmon? Salmon act like a ski lift that hauls nutrients out of the ocean's depths and delivers them thousands of miles inland. At sea, everything from orcas and swordfish to tuna, seals, and sharks preys upon salmon. But as salmon start migrating inland, they offer vast amounts of fat and protein to land-bound predators such as brown and black bears, bald eagles, wolves, foxes, ravens, gulls, and even ducks. What's even more surprising, though, is the way salmon benefit the *trees*.

"Predators feed on the majority of the body, but not the whole thing," explains Reid. "They leave behind a whole pile of ocean-derived nutrients."

As their bodies decay, salmon leach nitrogen and other nutrients into the soil, which are necessary for plant growth but can be really rare in certain ecosystems. So, when a bald eagle or bear drops that dead salmon deep in the woods, they're actually transporting molecules from the ocean into the lap of plants that have never and will never see the sea. Amazingly, scientists can even use the atomic weight of nitrogen isotopes to determine whether the atoms in a tree trunk originated from the land (or more specifically, the air, since 78 percent of Earth's air is made up of nitrogen) or from the sea. And this allows us to literally track this flow of life-sustaining elements and minerals from the darkest deep to the highest hills.

"You can see the signal in the trees themselves, which is why we call them salmon trees and salmon forests," says Reid. "You can detect these clear salmon signatures in the annual growth rings of trees."

Of course, people benefit from these animals, too.

For thousands of years, the Nisga'a First Nation has centered its calendar on when the salmon will arrive. Not only are the fish caught, dried, and otherwise preserved so that those critical calories can last the Nisga'a all year, but the salmon run is a keystone cultural event where everyone from elders to youth intermingle and whole families take part in the harvest.

"Most Indigenous cultures up and down the West Coast talk about themselves as Salmon People," says Reid. "I think it's out of that recognition that our long-term survival is deeply tied up in these fish."

By the way, if you eat salmon or other seafood, then you've got ocean-derived nitrogen isotopes in you, too. Which is just another reminder of the many ways in which we are connected to worlds we don't often think about but upon which life on this planet depends.

Firefly nymph
(*Photuris* sp.)

PART V

INTRIGUING INVERTEBRATES

THE VICIOUS AND DELICIOUS

BLUE CRAB

TYPE: Crustacean **SPECIES:** *Callinectes sapidus* **STATUS:** Not Evaluated **RANGE:** Atlantic Coast of the Americas, from Nova Scotia to Argentina **SIZE:** Up to 9 inches across; weighs up to 2 pounds **LIFESPAN:** Up to 4 years

---

CRABBERS IN CANADA, the United States, and Mexico haul hundreds of millions of blue crabs out of Atlantic waters each year, and those crustaceans wind up in crab bakes, crab cakes, crab dips, and crab bisques all over the continent. In the U.S. alone, blue crabs are worth around $200 million each year. But even if you know the sweet, delicate taste of crab meat, how many of us are familiar with the curious critter it comes from?

Blue crabs are hand-holdable crustaceans with brownish-green backs, creamy white underbellies, and brilliant cobalt-blue markings on their legs and claws. Females also sport a little dash of pink on the claw tips, almost like a dab of fancy nail polish. Of course, by the time a blue crab gets to our plates, it's red. The same thing happens to lobsters and shrimp, by the way, and this is actually due to a pretty neat chemical reaction.

Here's how it works: Lots of crustaceans have a red protein hiding in their shells known as astaxanthin. The thing is, another protein known as crustacyanin likes to wrap the astaxanthin in a biochemical bear hug, which hides it from our view while a crab is alive. However, when you apply heat to a blue crab's shell—either through grilling, steaming, boiling, or baking—the crustacyanin protein starts to fall apart, allowing the astaxanthin to burst forth.

In other words, add a little heat and bam! A blue crab turns red.

By the way, eating crabs is a pretty ancient practice. Evidence from a cave in Portugal shows that Neanderthals cooked brown crabs on hot coals and then broke the crustaceans open with tools as early as 90,000 years ago. And much more recently, on this side of the Atlantic Ocean, archaeological evidence from middens, or ancient trash heaps, reveals that Native Americans had been dining on Chesapeake Bay blue crabs for at least 3,200 years (though scientists have yet to uncover any evidence of Old Bay Seasoning).

Studying such things can teach us how blue crabs have changed over time. For instance, by measuring the tips of these preserved crabs' claws, scientists could determine that, back in the day, blue crabs were likely about twice as large as they are now. This means there were once blue crabs with shells stretching a full 10 inches in length, whereas today the species usually tops out at 5 inches. This is probably because we love blue crabs so much, and we're so good at harvesting them, that the decapods are rarely able to live long enough to get so massive (decapod means "10-footed" and refers to the order of crustaceans that includes crabs, lobsters, crayfish, shrimp, and prawns).

While humans and blue crabs are clearly linked by gastronomy, these animals thoroughly predate our own species. The fossil record suggests that blue crabs arose sometime during the Pleistocene epoch, which began 2.6 million years ago (*Homo sapiens* emerged roughly 300,000 years ago). What's more, the infraorder that blue crabs (aka Brachyurans or true crabs) belong to has been so successful across all that time that evolution keeps replicating it. That is, if you looked at all the various kinds of decapods and the evolutionary lineages they come from, you would find example after example of crustaceans that originally looked like lobsters or hermit crabs that have evolved over millions of years to look more like true crabs (the true crab shape is characterized by being short, generally round, and with a hidden or folded tail).

Neither porcelain crabs, hairy stone crabs, nor king crabs are actually closely related to the blue crab. Rather, they hail from other crustacean lineages that have slowly become more crablike over the ages.

The trend of evolution bending toward things that look like crabs is so pronounced, in fact, that scientists gave the process its own name—carcinization.

Beautiful, Savory Swimmers

The blue crab's scientific name is *Callinectes sapidus,* which is a combination of Latin and Greek. "It means, basically, 'beautiful swimming creature with a savory taste,'" says J. Sook Chung, a carcinologist (or crab scientist) at the University of Maryland Center for Environmental Science.

While many other crab species are short and squat, blue crabs have a flat, streamlined shape. Their last pair of legs has also been modified by evolution to look like paddles, and they use these appendages to dart through the water like race cars. Unlike land crabs, blues always have to be underwater, says Chung.

Chung is an expert on blue crabs, by the way, which makes her given name, Sook, pretty darn interesting, too. Because while a male blue crab is known as a Jimmy, and a young female crab is known as a Sally, a female crab mature enough to carry eggs is known as a Sook. Chung says her name is purely a coincidence. She is from South Korea, but the word "sook" actually comes from China. "So I joke that I was destined to study blue crabs," she says.

How do you tell Sooks apart from Jimmies? Look on their underbellies. All blue crabs have what's known as an apron, or a hardened plate near their nether regions. In females, this plate is large and dome-shaped with a little nub on top, not unlike the silhouette of the Capitol Building in Washington, D.C. But on males, this plate is noticeably smaller and pillar-shaped, like the Washington Monument.

The reason? Females open this apron to house their eggs, says Chung, so their plate has to be large enough to accommodate a brood that can be between two and eight million strong.

With so many crabby babies churned out from each female, you'd think North America's coastal waters would be positively overrun with swarms of blue and pink pinchers. But in 2022, scientists reported the lowest blue crab population numbers in the Chesapeake Bay since stock assessments began at the end of the 1980s. So what happened to all the bazillions of crabs that should be out there?

To answer that, you need to know about the blue crab's life cycle.

Sooks are exceptional mothers, says Chung. With their eggs nestled safely in their aprons, the females care for the developing embryos for two to three weeks, protecting them from all the hungry mouths of the sea and making sure they have plenty of fresh, oxygenated water wafting over them at all times. (By the way, when a Sook is with eggs, she is sometimes called a sponge crab, because of the appearance of the egg mass.)

But once the larvae hatch, they are on their own. And life can be tough at this stage, because the little squigglies are smaller than a pencil point. They can't even really swim, so they just go where the currents take them, eating whatever plankton will fit inside their mouths.

The larvae grow rapidly, molting their shells more than once a day. At this point, they don't look very much like a blue crab at all. Rather, they are microscopic monsters made up of see-through legs, bulging black eyeballs, and shrimp tails. After about a month to a month and a half, the larvae go through yet another molt, at which point they start to don that classic true crab shape. They can also swim now, and will use environmental cues such as the saltiness of the water and light levels to navigate the currents and arrive back at the estuaries they were born in.

"From eggs to the adult stage, they have to shed their shells 27 to 29 times," says Chung. All of this can happen between April and November in warmer waters such as the Gulf of Mexico. In the cooler waters of the Chesapeake Bay, the path to adulthood can last up to 18 months—for those that make it, that is. And not many blue crabs do.

Of the millions of eggs each Sook produces, just one to two will survive to adulthood. The rest disappear into the wild blue as food for innumerable creatures, from fish and oysters to jellies and whale sharks. Entire ecosystems depend on the annual blue crab fountain of life, though it's obviously a bit of a bummer for all those little larvae who never achieve their adult form.

Let's not get too precious about it, though. Blue crabs gobble each other up, too, in addition to preying upon clams, oysters, and mussels, which they crack open with their powerful claws. They'll also eat dead fish, plants, and basically anything that can't escape their pinchers.

Blue crabs are both predators and prey. Scavengers and cannibals. Nearly invisible larvae destined to die and highly sought after regional delicacies upon which countless humans depend.

As for Chung, she says she "has no desire to eat blue crabs." But she dissects them regularly and often only needs to keep the crustaceans' eyestalks for her experiments. Not wanting any of the life givers to go to waste, after the slicing and dicing is done, she sends out an email to her colleagues. "I have some blind crabs," it reads.

And everyone comes running.

THE WARRIOR QUEENS

CARPENTER ANTS

TYPE: Insect **GENUS:** *Camponotus* **STATUS:** Not Evaluated **RANGE:** Literally everywhere
SIZE: Up to half an inch long **LIFESPAN:** Less than a year for workers, up to 20 years for queens

---

WITH MORE THAN 14,000 SPECIES known to science and thousands more waiting to be described, ants are easily one of the most diverse and numerous life-forms on this planet. In fact, if you got every wild bird and mammal on Earth and put them on a scale, ants would outweigh them all—no contest.

From deserts and swamps to forests and plains, ants will be out there harvesting resources. On trees, beneath rocks, and furrowed within every rotting log, there will be ants waging war against their neighbors. Even in urban areas, where concrete and steel overwhelm soil and leaves, ants reside beside city stoops and in between sidewalk cracks, building six-legged civilizations of such surprising complexity that scientists still haven't discovered all their secrets.

"Just like how we build giant skyscrapers in cities, ants do it, too, but below-ground," says Corrie Moreau, an entomologist and evolutionary biologist at Cornell University. "In fact, they're more important than earthworms, because they create so much new and important habitat by building these giant fortresses."

But of all the ants in North America, none may be more prominent than the carpenters. Sometimes growing as large as a licorice-flavored jelly bean and often black or red (or reddish-black), the name "carpenter ant" represents any of hundreds of species within the genus *Camponotus*. (All carpenter ants belong to the genus *Camponotus,* but not all ants in the genus *Camponotus* are considered carpenter ants.)

"We sort of reserve 'carpenter ants' for the ones that get into wood and actually expand galleries," says Moreau, who is also a National Geographic Explorer.

Carpenter ants do not eat wood, like termites, but rather remove it mouthful by mouthful until the space left behind is big enough to travel through or live in. This is why little piles of sawdust scattered around the house can be evidence of a carpenter ant infestation—they are literally carving away at your infrastructure. And while I hope reading this book has provided you with a general appreciation for all creatures great and small, even Moreau says there are limits to the live-and-let-live philosophy.

"It's one of the few groups of ants where I say, if you have them in your house, you really do want to call an exterminator," she says.

The Hidden World Below Our Feet

Did you know that pretty much every ant you've ever seen was a female? This is because male ants don't forage for food, nor do they take care of the young or even defend the nest. In fact, you could crack open a carpenter ant colony right now and odds are you'd find precisely zero males, because male ants don't even exist most of the time. The queen produces males just once each year for a single, short purpose.

"They're just flying buckets of sperm," says Moreau.

How do you know if an ant is male? The easiest way, says Moreau, is to see if the ant has wings. If so, there's about a 50 percent chance it's a male.

When it comes time to reproduce, the queen lays a special batch of eggs that will grow into ants completely different from any others in the colony. In this brood of both males and females, the ants will develop wings which they use to flap out into the world and meet ants from other colonies, which they will mate with. This ensures that new colonies benefit from genetic diversity, which helps species buffer themselves against total annihilation by things like disease, changes to habitat, or predators.

For the male carpenter ants, death comes quickly after the transaction. But females have a different fate.

Once mated, the new queen-to-be will hit the ground, rip off her own wings, and burrow into the earth, where she lays a small clutch of eggs. The hatchlings, known as nanitics, will be a smaller, weaker, and paler batch of workers than all the ants that come after them, and in the beginning, the queen feeds the nanitics with kisses. Or, more specifically, with nutrients produced by her salivary gland.

In time, the nanitics develop enough strength to take over the daily rigmarole of colony life: primarily, searching for food and taking care of the queen as she lays the second batch of eggs, which will turn into proper, full-size workers known as minors. If the colony survives long enough, the queen will start laying eggs that turn into even bigger supersoldier ants with weaponized jaws, known as majors. In just a few years, carpenter ant colonies can scale up from just a handful of nanitics to bustling metropolises some 100,000 ants strong. While some carpenter ant species house multiple queens in a nest, lots of these societies are built by that first little egg layer—and if she dies, the whole kit and caboodle eventually go with her.

"Everybody thinks it's best to be the queen in an ant colony. But it's the worst job," says Moreau. "You never see the sunlight again. You don't care for the nest. You don't get to do any of the fun stuff. You are just an egg-laying machine."

Worker ants, on the other hand, get to cycle through different roles as they grow. "This is known as age polyethism," says Moreau.

Like monarch butterflies, ants begin life as eggs before hatching and spending a few weeks as helpless, maggotlike eating machines called larvae. "They're just a sack with a mouth," says Moreau. In fact, you've probably seen these little buggers getting carried around by other ants if you've ever lifted a rock and caused the mass exodus of a hidden ant colony. After growing in size and molting a few times, the larvae pupate or transform into adults. And then it's time for the kiddies to go to work, rotating among roles as nurses to the next generation of larvae, tunnel cleaners, nest builders, or delivery girls whose only job is to grab food dropped off by the foraging crew.

Only the oldest ants in the colony get to go outside and hunt for food or do battle with other ants. And because all the ants we're talking about here are female, this means every ant you've ever seen patrolling a parking lot or dragging the hulking carcass of a bumblebee was actually a tiny, old, steely-eyed grandma.

Actually, since only the queen reproduces, they're more like great-aunts. Great-aunt ants.

Carpenter ants have also been found to incorporate other insects into their workflow. For example, carpenter ants share many habitats with another, smaller insect known as an aphid. Aphids have a straw for a mouth, which they use to suck the juices out of plants. "But it's kind of like if we took a straw and stuck it into a fire hose," says Moreau. "Our cheeks would explode."

In other words, the force of the plant's juice flow greatly exceeds the aphids' ability to drink, so the insects have evolved a digestive system than can shunt excess liquid around the stomach and straight out the aphid's backside whenever necessary. In turn, ants evolved a symbiotic relationship with many aphid species where they protect the little suckers against parasitic wasps and predatory ladybugs in exchange for access to those sweet butt juices.

Moreau says some ant species even herd the aphids, like we do with livestock, driving them down to the base of the plant and tucking them in each night before waking them up the next morning and directing them up to parts of the plant they haven't grazed yet.

All of this is probably happening in your backyard as we speak, by the way. Not to mention the ant zombies.

Before you start boarding up your doors and windows, let me explain: In the forests of South Carolina, Florida, Missouri, and Georgia, scientists have discovered evidence of a fungi called *Ophiocordyceps.* This parasite invades the brains of individual carpenter ants and tricks them into making a death march out of the colony and into the forest, like brainwashed ant-omatons. Once there, the infected ant climbs high onto a plant and bites down on a leaf vein, at which point the fungus spreads through the ant's skull and sprouts a long stalk out of the insect's forehead, finally killing the ant in the process. The creepy growth is actually the fungi's fruiting body, and when it's ripe, it rains spores down onto the unsuspecting ants below, beginning the process anew.

If you could pick one animal to follow around for the rest of your life, the humble ant would keep you more entertained than most. They are warriors, explorers, ranchers, builders-of-cities, and sometimes, victims of zombie-making fungi.

"It's one of the things I love most about insects. You can live anywhere, and you can study wildlife," says Moreau. "You're actually watching biology in real time."

THE PINCHY ONES

CRAYFISH

TYPE: Crustacean **SPECIES:** More than 500 species in the infraorder Astacidea **STATUS:** Varies by species, from Least Concern to Extinct **RANGE:** Worldwide and throughout North America **SIZE:** Up to 9 inches long; weighs up to 8 pounds **LIFESPAN:** More than 10 years in some species

---

IF YOU ROUNDED UP every cat species on Earth—all the tigers, lions, leopards, and lynx—and then added every known species of canid—dogs, coyotes, wolves, jackals, and the like—that menagerie would still be a few dozen species short compared to the number of species of crayfish that live in, say, Tennessee.

Or Alabama, in fact. Go ahead and take your pick, because while Antarctica does penguins, and Madagascar has a lock on lemurs, the United States goes hard on crayfish. Of the nearly 700 species of crayfish known to exist on Earth, more than 500 of them call North America home.

"The southeastern U.S. is truly *the* biodiversity hot spot for crayfishes in the world," says Bronwyn W. Williams, research curator of non-molluscan invertebrates at the North Carolina Museum of Natural Sciences.

For those that may be unfamiliar with crayfish, the easiest way to think about these animals is that they're freshwater cousins to the lobster. And as you might expect for so many species, there is great variation with regard to crayfish size, shape, and color. For instance, longpincered crayfish *(Faxonius longidigitus)* found in the American Midwest and Southeast regularly grow up to six inches in length, as long as the U.S. dollar. And while that is pretty impressive, across the Pacific, there's a crayfish that can weigh an absolutely unbelievable 13 pounds, or about twice the weight of a Chihuahua. Known as the Tasmanian giant freshwater crayfish *(Astacopsis gouldi),* this species lives to an astounding 60 years of age.

In the American South, these crustaceans are referred to as crawfish. In the Midwest, they're crawdads. And up north, folks usually say crayfish. But that's just a few of the lovely little nicknames applied to these creatures. Others include crawdaddies, baybugs, crabfish, freshwater lobsters, mountain lobsters, mudbugs, rock lobsters, signal crawfish, craycrabs, crawcrabs, and yabbies. That last one is from Australia, mind you, but personally I think yabbies is just fun to say.

"It's very regional," says Williams, who is Canadian but grew up in western New York, and as such, says crayfish.

Most people think of crayfish as living in streams, which they do, but they can also survive in lakes, ponds, rivers, and puddles. Some species do just dandy in those little ditches beside the highway.

As decapod crustaceans, crayfish have 10 legs, but the frontmost pair has evolved into large claws, also known as chelae. Crawdaddy heads come to an adorable little point, known as the rostrum, with beady eyes on either side. On their tails or abdomens, crayfish also have yet another five pairs of tiny minilimbs known as swimmerets. These little nubs actually generate thrust and allow the animals to scooch backward rapidly in order to escape threats, like a big grubby human hand trying to grab 'em.

While the stereotypical crayfish color is brown, these crustaceans can come in a rainbow of hues depending on the species. "You can have crayfishes that are bright red or bright orange or bright blue," says Williams.

The red swamp crayfish *(Procambarus clarkii),* which has a native distribution from Oklahoma to Illinois and Alabama, has claws that look like they're made out of ruby. The Monongahela blue crayfish *(Cambarus monongalensis)* of Ohio, Pennsylvania, and West Virginia is so blue that it looks like its armor was dipped in liquid cobalt. And plenty of crayfish, like the Benton County cave crayfish *(Cambarus aculabrum)* of Arkansas and Missouri, have lost their colors entirely, and now walk around like translucent ghosts in their subterranean lairs.

Oh, and while we're talking about colors, did you know crayfish blood is actually blue? This is because it contains an oxygen-transporting protein known as hemocyanin, and hemocyanin has copper in it, which reflects blue light. Whereas our blood has hemoglobin, which contains iron, and reflects red light.

Lobsters of the Land

For an expert on animals commonly thought of as needing water to survive, Williams spends a surprising amount of time digging in the dirt, often in the middle of fields. This is because some crawdads are actually landlubbers.

According to Williams, all crayfish do a bit of burrowing. For most species, this might simply be excavating a little space beneath a rock at the bottom of a creek. These are known as tertiary burrowers.

Next, you have secondary burrowers, and these are the crayfish responsible for all those coin-size tunnels lining the muddy banks of streams. (Growing up, I'd always assumed snakes made these, but it turns out my fears and fascinations were extremely misplaced.) These yabbies may venture out and forage in the watery domain, but their happy place is down in the hole, which leads to a watery pit where they can avoid predators.

Finally there are the primary burrowers, and these crayfish are the weirdos of the lot. Rather than huddling under a rock or ball of roots below the waterline, these mountain lobsters take to the land—burrowing deep into the soil and creating neat little hovels complete with mud chimneys, secret escape tunnels, and indoor pools that tap into the water table. Burrowing crayfish will also plug their chimneys when the world above gets too dry, trapping moisture within.

Like all large crustaceans, burrowing crayfish breathe using gills. But unlike creatures such as the blue crab, which can only breathe while underwater, crayfish gills can wick oxygen out of the air, so long as they remain wet. And this allows crayfish to walk between worlds.

Williams says she's found crayfish thriving at least a quarter mile away from any obvious water source. But remember, just because we can't see water doesn't mean it isn't there. "Where's that groundwater?" says Williams. "That's where you're going to find the crayfish."

In her quest to take stock of crayfish biodiversity in North America, Williams has excavated crayfish burrows that plunged five feet deep into the earth. And that's not where the tunnel ended, she laughs, just where she gave up and stopped digging. A colleague of hers has gone eight feet down and found nothing but the black abyss of more crayfish burrows waiting for him at the bottom.

Crayfish have tons of natural predators, such as great blue herons, raccoons, otters, foxes, snakes, turtles, fish, and hellbenders. In turn, crayfish are omnivores who scuttle through aquatic ecosystems and mow down everything from plants, detritus, and insects to fish eggs, tadpoles, and rotting flesh.

"Crayfish are a central component of food webs," says Williams. "We joke around that crayfish eat everything and everything eats crayfish."

While it's generally not a good idea to move plant or animal species around, it should be noted that some golf courses have used non-native red swamp crayfish to control the weeds in their water features. And in at least one study, researchers noted that the same crawdad species could actually reduce the incidence of schistosomiasis in Kenyan schoolchildren. This is because the crayfish chow down upon the snails that carry the parasitic worms that cause the disease in people, thus breaking the pathway to infection.

But wait—crayfish are also ecosystems unto themselves!

"Think of a crayfish like a rock in a stream, right? Whatever settles on that rock is also settling on that crayfish," says Williams.

And all that pond scum and river riffraff feeds entire species of crayfish worms and ostracods—which are even tinier crustaceans—that graze the crayfish's shell like herds of cattle. Mostly, the hangers-on do the mudbugs no harm, but if the host gets a wound, the entourage takes the opportunity to sip some of that protein-rich blue blood.

"I've collected six different species of worms and two different species of ostracods on a single crayfish," says Williams, who says each species stakes out its own territory on the crawdad—some on the legs, others on the rostrum, and still more in the gills, leg crevices, or tail.

"Crayfishes are landscapes for entire communities of critters that most people don't even know exist," she says.

By the way, Williams says it's absolutely okay for you to go out and experience these walking fortresses of biodiversity for yourself—so long as you're not worried about getting pinched! Just be gentle and respectful. Put the animals (and the rocks that hide them) back where you found them. Oh, and with regard to burrowing crayfish, remember:

"If the burrow is not in use, or has been abandoned, it could be occupied by organisms you might not expect or want to encounter," says Williams. "You don't want to blindly stick your hand down any hole in the ground."

Frankly, that's just good life advice.

THE SECRET SAVAGES

FIREFLIES

TYPE: Insect **FAMILY:** Lampyridae **STATUS:** Varies among family, from Least Concern to Critically Endangered **RANGE:** Worldwide; in North America, from Alaska to Nova Scotia to Chiapas **SIZE:** Up to 1 inch long **LIFESPAN:** Up to 4 years

---

DID YOU KNOW that there are around 2,000 species of firefly on this planet, and that North America is home to more than 120 of the buggers? Or that scientists believe fireflies evolved their flash as a way to ward off nocturnal predators, but that, somewhere way back in time, that signal was co-opted by certain species as a way to flirt?

Today, all firefly species can be broken down into three categories: You have the lightning-bug fireflies (the ones that fly and glow as a way to attract mates), the glow-worm fireflies (the ones where only females glow to attract males and usually only do so from the ground), and the dark fireflies (the ones that don't glow at all but rather use scents to find each other).

Amazingly, firefly lamps come in a whole spectrum of colors, from pale blue (genus *Phausis*) to neon green (genus *Photuris*) to bright yellow (genus *Photinus*) and even burnt orange (genus *Pyractomena*).

And the more scientists study these bioluminescent beetles, the more they have discovered that every species of glowing firefly has its own unique flash language that helps tell it apart from other species! This means that on a single summer's eve, a field of flickering phantoms could actually include more than a dozen different kinds of fireflies, each tapping out a visual Morse code in the hopes of getting lucky in the dark.

Some species like to flash in the treetops, while others prefer to blink just a foot or two above ground level. Some fireflies only flash when the temperature is 80 degrees, or 71 degrees, or 67 degrees, while others have no trouble lighting up as long as it's 50 degrees or warmer. One species might flash six times in a row and go dark, while another keeps its lamp on for a full 15 seconds as it hovers in a straight line. One common species, known as the big dipper firefly, scrawls out a distinct J shape every time it turns on. (By the way, with all the intricacy on display here, you can help lightning bugs find each other and mate simply by turning off all your exterior lights at night—a good idea, since these creatures are generally in precipitous decline.)

Consider also that while we take firefly flashes for granted as things that just, well, happen, the beetles can only produce their light show because of a chemical reaction among an enzyme known as luciferase, a compound called luciferin, another incredibly important compound known as adenosine triphosphate (ATP), and everybody's favorite element, oxygen! Even more spectacular, the beetles can control that chemical reaction, using special chambers that mix all those potions together at a moment's notice and create hardly any heat, which might otherwise cook the critters in their shells.

But as magical as lightning bugs are as adults, there is another side to the animals that I think is well worth exploring. Because beneath that facade of quiet peacefulness, fireflies are secret savages.

The Lightning Bug's True Form

"People see those beautiful, ethereal insects that are out there in the field, lighting up the night, and they get this very warm, romantic feeling," says Sara Lewis, a biologist at Tufts University who has been studying fireflies for around 30 years. "But they don't really appreciate the backstory."

The truth is, nearly everything we know about fireflies happens in the span of about two weeks—the time the beetles spend as adults. But the vast majority of the firefly's lifespan plays out over the two to three *years* that they spend as wingless larvae. So the fact that we focus on the adult beetles as the be-all and end-all of firefly existence is kind of like if an alien zoomed in on some people honeymooning in Tahiti and concluded that all humans do is lie on the beach and sip beverages out of coconuts (which sounds pretty awesome, but of course does not accurately describe the human condition).

If you really want to get to know the firefly, says Lewis, then we'll need to venture underground.

All fireflies begin their lives as eggs so small that they could balance on the edge of a credit card. Females lay those eggs on moist soil, and the eggs of some species will even glow softly when disturbed, likely as a way to discourage other animals from eating them for breakfast. Once the eggs hatch though, the violence begins.

"Because most fireflies in North America spend their childhood underground, we don't normally see them," says Lewis, who is also the author of *Silent Sparks: The Wondrous World of Fireflies*. "But they're down there, and they're using their very, very sharp, sickle-shaped jaws to gorge themselves on earthworms."

Scientists aren't entirely sure how the larvae catch wind of the worms they hunt, but it probably has something to do with following the chemical trails the invertebrates leave behind, says Lewis. In any case, we do know what happens once the target is locked.

"They just start biting and biting and biting," she says. "We think they are injecting the earthworm with some kind of toxin that paralyzes its muscles."

Armed with its chemical weapons, a single firefly larva can take down an earthworm many times its size. But the tiny marauders have also been shown to work together, almost like a wolf pack.

"It's a little bit like the worm-riders in *Dune*," says Lewis, referencing the popular sci-fi novel and movie franchise. "They just glom on to the earthworm, and even if it is struggling and thrashing, they just hold on and keep injecting it with that paralyzing neurotoxin."

Eventually, the earthworm succumbs. But it does not die. Rather, its heart continues to beat, keeping its innards fresh for the firefly larvae, which take their time carving the beast apart and devouring it, in steak-like segments, over the course of several days.

“They’ve got jaws but no teeth, so they secrete enzymes to liquefy their prey,” says Lewis. “Then they kind of just suck up the earthworm like a smoothie.” As it turns out, earthworm smoothie is a wildly effective way to pack on weight, which the larvae need so that they can eventually undergo metamorphosis and turn into flickering adults.

After a big kill, the firefly larvae become so plump, Lewis says, that their legs can’t even touch the ground anymore, and they all just lie on their backs for a few days as their bodies digest the buffet. “And then they go back to eating the earthworm, which is still there, because it’s been paralyzed,” says Lewis.

While firefly larvae all over the world practice some variation of this carnage, North America is also home to the world’s only firefly that has a predatory lifestyle as an adult.

Remember how a firefly’s flash is a secret code? It helps members of the same species find each other in the night while avoiding the waste of energy (and embarrassment!) that might come from trying to mate with the wrong species. Well, female fireflies of the *Photuris versicolor* complex have learned that by mimicking other species’ flashes, they can lure male fireflies to their grave.

This trick has earned these fireflies a lovely little nickname: femmes fatales.

Amped up on hormones and perhaps the knowledge that their mortality is fleeting, the fellas never see it coming. All they know is a female winked at them somewhere out in the darkness. And like many fools before them, they soar headlong into the night expecting to find a trail of rose petals. Instead, they meet their doom.

After the femmes fatales trick a male of another species into coming in close, the females pounce out of the shadows and devour them. I wish I could tell you that the female fireflies were tiny vigilantes out to destroy the patriarchy. But there seems to be a different endgame at work.

Scientists believe that the ladies turn predator because their species lacks the ability to produce lucibufagins, or defensive compounds that make lightning bugs taste nasty to birds and bats. Males of other firefly species make them, though, both as a way to ward off predators and as a way to woo females of their own species. (The males offer their females a neat little care package of lucibufagins and other nutrients, like the firefly equivalent of a box of chocolates.)

But by gobbling up those males instead, the femmes fatales fireflies appear to be able to steal their cousins' weapons and infuse them into their own eggs, thereby ensuring the success of the next generation.

So the next time you see lightning bugs quietly flashing across a summer field, remember that each and every one of them started life as a ravenous little hellhound. And now, all of them are betting on a soft glow in the darkness. For some, the night will end in love. For others, blood.

But all of it is magic.

THE RIVERKEEPERS

FRESHWATER MUSSELS

TYPE: Mollusk **SPECIES:** More than 300 species in North America **STATUS:** Varies among order, from Least Concern to Extinct **RANGE:** From the Arctic Circle to the Yucatan Peninsula **SIZE:** Just over an inch to 11 inches long; weighs up to 5 pounds **LIFESPAN:** More than 150 years

---

RIGHT NOW, as you read this, there's an ancient being lurking on a river bottom nearby. It may have been sitting in the same spot for more than a hundred years. It probably has a silly name, such as the Orangefoot pimpleback, Appalachian monkeyface, shiny pigtoe, Tennessee heelsplitter, rough rabbitsfoot, or the purple cat's paw. At first glance, you could be forgiven for thinking it looks like little more than a glob of snot hiding inside a two-piece shell.

And for too long, we've taken these globs of snot for granted.

Mussels are mollusks related to clams and oysters. While reminiscent of living rocks, these incredibly important animals help anchor creek- and riverbeds where they purify water by straining it for tiny bits of food.

"They eat diatoms, algae, and really fine particulate matter," says Rachel Mair, a mussel biologist and project leader with the U.S. Fish and Wildlife Service. "They filter everything and then decide what they want to keep. And then they package all of that waste material in mucus and discard it."

Now, that may sound gross, but for aquatic insects, mussel mucus packets might as well be dumplings. (And it's ideal to have a healthy aquatic insect population, because insects feed the fish we like to catch and contribute to a functioning ecosystem.)

At the same time, by removing literal tons of sediment from rivers and streams, Mair says it's "kind of [the mussels'] way of recycling that water and making it usable." A single mussel can process around 15 gallons of water each day, removing not only sediment but also nitrogen, toxins, and heavy metals. No wonder mussels are sometimes called the river's liver.

Now here's where things get freaky. Freshwater mussels can't move. Most species spend their entire lives cemented to the same underwater nook. At the same time, freshwater mussels have an absolutely enormous range, inhabiting creeks, streams, and rivers essentially everywhere in North America. So how does a sedentary ball of goo conquer new lands?

Well, by deception.

When a female mussel is ready to reproduce, she must first lure a fish down to the creek bottom so she can squirt it in the face with thousands upon thousands of her microscopic larvae. And each species has evolved a different strategy.

Some mussels, such as the giant floater, spin nets of mucus for fish to swim through. And when the fish come out the other side, they've already become parasitized. Other mussels, like the Higgins eye pearlymussel, have skin flaps decorated with stripes and eyespots that look remarkably like darters, minnows, and other species of small fish that larger predators love to eat. Other species, such as the western fanshell, dangle long strings of mucus that look like wriggling worms. The rainbow mussel may be most convincing of all. It's evolved a mantle flap that looks for all the world like a crayfish. What's more, when a smallmouth bass swims near, the mussel can wag its skin and make the crayfish decoy move. This is nature's equivalent of the dancing lady tattoo.

One absolutely diabolical mussel genus called *Epioblasma* draws fish in with a lure and then clamps closed on their face like a bear trap. Some of these species have even evolved tiny "teeth," or denticles, on the edges of their shells, the better to grip their victims and bend them to their will.

Whatever the species-specific ruse, the endgame is the same. The freshwater mussel douses the fish with a cloud of parasitic spawn, which get swept up into the fish's gills. Once there, the baby mussels snap closed around bits of tissue and hang on for dear life. Over time, the itty-bitty vampires leach nutrients off the fish. Then, after the larvae have matured, they will drop off their mobile nursery in a new, productive part of the river system. Once settled, the mussels anchor themselves to the floor and begin the cycle anew.

Pretty dramatic life cycle for a stationary invertebrate, right? Unfortunately, what has worked for these animals since the Triassic Period has now become part of their undoing. Because though the U.S. has around 300 species of these marvelous mollusks, more than 70 percent are classified as endangered or possibly extinct.

Scientists are still trying to pinpoint the mussels' largest threats, but most point to water pollution, infectious diseases, and climate change. The answer likely depends on which species and watershed you look at, but there's one thing mussels have in common: They're the most endangered group of wildlife in the United States.

Bathed in Blood

In a small facility on the frozen banks of Ohio's Scioto River, several thousand freshwater mussel larvae pulse against each other in a petri dish full of rabbit blood. While this may sound like the work of a mad scientist, it's actually the latest and greatest technology in the fight to keep freshwater mussels from disappearing forever.

Though we have only just begun to take notice, mussels and fish have been doing this strange evolutionary dance for hundreds of millions of years. And over all that time, mussels have gotten better and better at parasitizing their hosts. But optimization comes at a cost. For all the brilliance and subterfuge present in the lures listed earlier, each can only work on a few or maybe even just one species of fish.

"The parasite becomes so specialized in evading the host's immune system that it gets to the point where it really can't parasitize anything else," I learned from G. Thomas Watters, an Ohio-based malacologist, or a scientist who studies mussels. (A sad note: Dr. Watters passed away in 2019. This chapter is dedicated to his memory and passion for freshwater mussel conservation.)

At the same time, humans have spent the past 300 years damming up this continent's rivers, in part to allow ships and barges to go places they couldn't before and in part to generate hydroelectricity.

In some cases, mussel species simply died off when their rivers were dammed up because they couldn't handle the change from fast-flowing water to what is essentially a standing lake. But in other cases, scientists have watched populations of mussels soldier on for decades without ever producing a single offspring.

Why? Because all the host fish have been cut off from those same dams. And in some places, the fish that mussels require may have even gone extinct.

Watters spent the last 35 years of his life trying to save these bivalves. As the curator of mollusks at The Ohio State University, he oversaw the largest freshwater mussel collection on the planet, some 1.6 million specimens in all. "That's where mussels go when they die," said Watters, referring to the countless drawers, jars, and boxes' worth of shells in the collection. While many species at OSU are extinct, drive north just a half-hour to find the Columbus Zoo's Freshwater Mussel Conservation and Research Center, where Watters once served as science director. Here, the focus is on the living.

Once a popular 1980s wedding venue, this converted facility now houses around 700 mature mussels representing 43 species from around the region. Where brides and grooms performed their first dances, there are now huge vats of mud, sand, and murky water, each supplied by a labyrinth of pipes that lead out to the Scioto River just a stone's throw from the back door. In the kitchen, where chefs once prepared shrimp cocktails and Chilean sea bass for their guests, there are now rows upon rows of tiny plastic fish tanks, each with a different species of darter, trout, or grass pickerel. With any luck, most of them are infected with baby mussels.

By relentless trial and error, mussel scientists all across the U.S. aren't just learning how to keep mussels alive in captivity, but they're also capturing and nurturing various species of fish in order to discover which species can be infected by which mussel.

And that brings us back to rabbit blood.

Until a host species can be found, blood can serve as the basis for an advanced freshwater mussel propagation technique known as in vitro. Instead of getting their safety and sustenance from a host fish, the in vitro method coaxes mussel larvae to grow and transform into juveniles while bathed in a soup of rabbit blood, antibiotics, and other nutrients. Why rabbits? Because fish blood is almost impossible to purchase in sufficient quantities.

With in vitro allowing scientists to leapfrog an entire stage of the mussels' life cycle and crank out tons of captive-raised mussels in the process—mussels that can then be reintroduced to the wild—hope remains for many species on the brink. But in a freshwater ecosystem, the reaper's scythe remains ever overhead.

"There are so many outside influences," says Mair. "If there's a spill, mussels can't get out of the way," she says. "They're just kind of stuck with it."

Many sources of pollution aren't even illegal. "You have factories that have discharge permits that are allowed, by law, to release a certain amount of their chemical discharge into the stream. That's out of your control."

In 2021, the U.S. declared eight more species of freshwater mussel extinct. And as the riverkeepers disappear, so do all the ecosystem services they provide for free. Cleaning water. Buffering against erosion. Providing food for insects, which in turn feed fish, turtles, and bats.

"There are species that I've worked with that probably won't be here in my lifetime," says Mair.

THE ALL-TERRAIN ARACHNIDS

HARVESTERS (DADDY LONGLEGS)

TYPE: Arachnid **ORDER:** Opiliones **STATUS:** Vulnerable **RANGE:** Worldwide (except Antarctica) **SIZE:** Body is a half-inch long, while legspan can exceed 6 inches across **LIFESPAN:** 1 year

---

THERE'S A PRETTY PERVASIVE MYTH that goes something like this: "Daddy longlegs are the most venomous spider in the world, but their fangs are too small to pierce human skin."

Sound familiar? Well, this little nugget of conventional "wisdom" is not just incorrect. It's spectacularly wrong on three separate counts.

Daddy longlegs are arachnids that have tiny, pill-shaped torsos and legs that go on for days. They have a pair of shoddy eyes, breathing tubes known as trachea, guts for digesting, and even secret, trap-door genitals, but really, for the most part, these animals are just an absolute mess of legs and the chunky thing those legs connect to, like bicycle spokes to a hub.

Despite their appearance, daddy longlegs are not spiders at all. Like mites, ticks, and scorpions, daddy longlegs are spider cousins. Which is to say, daddy longlegs are arachnids that belong to the order Opiliones, not Araneae (where the spiders are classified).

Furthermore, spiders usually have eight eyes, while daddy longlegs usually only have two, each of which is rudimentary and can only see light and dark but probably not the look on your face when you spot it on the wall of a port-a-potty and run away screaming. Likewise, if you look at the animals from above, spider bodies have two main parts—the cephalothorax and the abdomen, with a waist in the middle. Daddy longlegs bodies are usually scrunched together into one part.

"Spiders look like they're wearing a little corset. Opiliones just let it all hang out," quips Mercedes Burns, an evolutionary biologist and Opiliones expert at the University of Maryland at Baltimore County.

But perhaps the easiest way to tell a longlegs from a spider? "If it's in a web, it's a spider, because Opiliones can't make silk," says Burns.

Getting back to the "most venomous spider on Earth" bit, well, you can stop worrying about daddy longlegs bites, because these creepy-crawlies don't even have fangs. "Their mouthparts look like little scissors," says Burns. "And they use them to cut off little pieces of food and put them into their mouth."

That bit about the potency of daddy longlegs venom? It's utter hogwash. Because daddy longlegs don't have venom. Not even a little bit.

Add it all up, and I'm telling you that daddy longlegs are physically incapable of causing you harm. You could dive into a swimming pool filled with the things and come out without a scratch. Though, with that many wispy legs, you might have to worry about death-by-tickle.

Speaking of those legs—a closer look reveals that daddy longlegs patellae, or knees, sit way up high in the air, suspending their body above the ground in a way that gives them an extremely low center of gravity, says Burns. This allows the arachnids to remain stable as they bounce around on legs that are less like rigid stilts and more like bungee cords. Below the knee, each leg has something like 18 to 20 tarsal segments, all of which bend and allow the leg to grab onto everything from a tree branch to a blade of grass. "It's almost like if your feet could wrap around things," she says. And, you know, if you had eight of them instead of two.

"Their second pair of legs have lots of sensory structures on them. And they use them to kind of tap around as they're walking," says Burns. "They can walk on those legs, too, but they don't prefer to because they use them to feel their way around, maybe even tasting and smelling the surfaces that they're coming into contact with."

Sometimes, the legs end in hooks, the better to grapple with, and some species even produce a kind of glue on their appendages that they use to catch prey in wet places, such as waterfalls. "I kind of think of them like a jack-of-all-trades," says Burns. "They're an all-terrain arachnid."

Now, as anyone who has ever tried to squash a daddy longlegs knows, their legs fall off their bodies so willingly, it's almost as if the arachnids want to be rid of them. But leg loss is a feature, not a bug.

"We regularly find Opiliones out in the environment that have lost legs," says Burns. "And they're still able to locomote just as well as individuals that have all eight." In fact, she says they can get by with as few as four legs, which means that even if they don't have the nine lives that cats supposedly do, they at least have four. Because lost legs are do-overs.

When a predator (or curious little kid) gets hold of a daddy longlegs, it's likely to focus its attack on one of the arachnid's legs. And since the poor critter doesn't really have any weapons to defend itself with, it's left with a choice—stay and die or cut its losses and run. This is why Opiliones legs are so incredibly easy to detach. They're basically designed to snap off and do so cleanly, lest the arachnid bleed out in the process. Once plucked, the leg also jumps around as the nerves continue to fire—a potential leave-behind distraction that occupies the predator while the now-slightly-less-leggy one makes its escape. Scientists call this leg-jettisoning ability autotomy.

Again, all of this seems wildly counterproductive to survival, but it's worked for the Opiliones for hundreds of millions of years. In fact, while we tend to point to animals like cheetahs, falcons, and blue whales when discussing the pinnacles of evolution, there's a hell of a case to be made that the supremely strange arrangement put forth by these arachnids may just be one of the most successful body forms on Earth.

"These things have existed for a really long time," says Burns. In fact, the first Opiliones fossils hail from 405 million years ago, which makes the order one of the oldest terrestrial animal groups. "But their bodies are more or less unchanged since then," says Burns. "I don't ever want to call something a 'living fossil,' because if it's alive today, it's still evolving. But it's an indication to us that their body is pretty robust to evolutionary and environmental change."

In other words, evolution looked at the daddy longlegs and said, "If it ain't broke, don't fix it."

What's in a Name?

Before we go any further, I should tell you that most experts don't like the name daddy longlegs. After all, not all daddy longlegs are daddies. Just like not all ladybugs are ladies.

The phrase is also highly localized, since the eastern United States and Canada is a hot spot for Opiliones biodiversity, with around 35 long-legged species. But over on the West Coast, harvester species dwindle, and folks tend to use the name daddy longlegs to refer to a species of cellar spider. In other words, using "daddy longlegs" as a descriptor is already pretty problematic, not to mention weirdly kinky. So, what the heck should you and I call these animals?

"A lot of us in the biz are calling them harvesters," says Burns, due to the perception that the animals are most commonly sighted in the autumn. And there's a reason for that—like petunias and geraniums, harvesters are annuals.

Come October, November, and December, all long-legged harvesters on the continent will die, hopefully not before mating and tucking some eggs into the leaf litter. In the spring, those eggs will hatch and give rise to the next generation of harvesters. "Harvesters have what's known as direct development," says Burns, "meaning that when they hatch out of an egg, they already look like a miniature adult." You're unlikely to ever see a baby harvester, though, because they're tiny and spend their first few molts (or growth phases) hidden on the ground or underneath tree bark.

Throughout the year, harvesters are, well, harvesting. "They're often eating things that are dead or dying," says Burns. And this can include both plants and animals. "Their poops go back into the ground and become dirt," she says. "They're recycling resources back into the earth."

By the time fall comes around, harvesters are looking to shed their exoskeletons for the last time, and this is why they often crawl up on window screens and garage walls. "They like to be upside down, if possible, because then they can push themselves out of their skin, using gravity to help," says Burns.

That, or they're looking for mates. Or just trying not to get torn limb from limb. "You could say Opiliones live short lives that are full of adventure," says Burns, ever the optimist.

So, the next time you see a harvester feeling its way along a wall or ceiling, remember that it means you no ill will. It's spent most of the year staying out of your way and recycling nutrients, but now instinct and the turning of the seasons have driven this curious creature up out of the grass and into your line of vision—a very dangerous place to be, indeed.

Despite what you've been told, harvesters are more defenseless than bunnies, mice, or doves. They are humble, bumbling beings at heart, and they'd sooner cut off their own legs than do you harm.

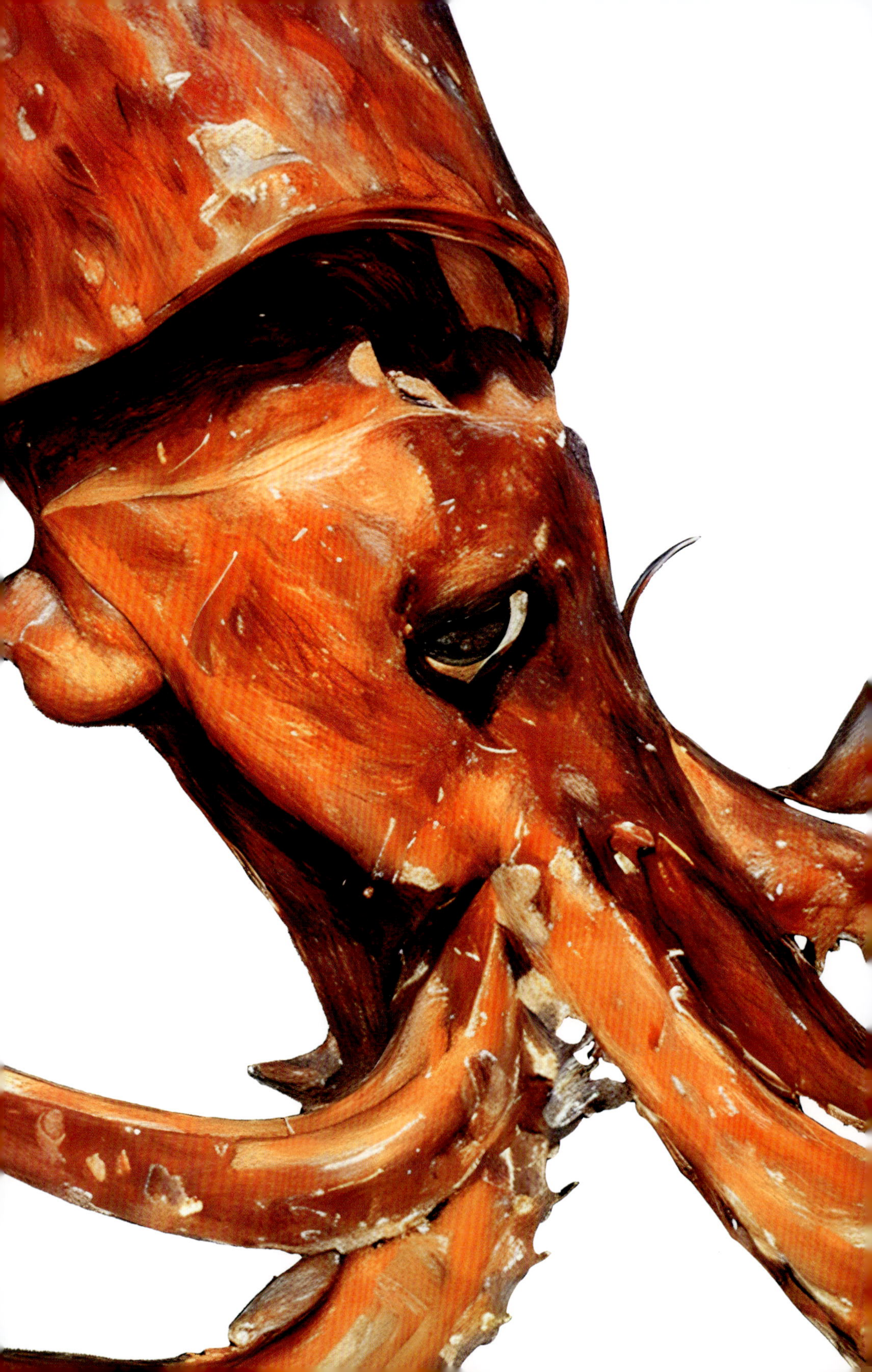

THE KRAKEN

HUMBOLDT SQUID

TYPE: Cephalopod **SPECIES:** *Dosidicus gigas* **STATUS:** Data Deficient **RANGE:** From South America's coastline to Alaska **SIZE:** Up to 8 feet 3 inches long; up to 110 pounds **LIFESPAN:** 1 to 2 years

---

STAND ON THE WEST COAST of North America and look out over the waves, and visions of ocean wildlife might start dancing through your imagination. Pods of bottlenose dolphins and humpback whales, flocks of gulls, and processions of pelicans. Sea stars, sand dollars, leatherback sea turtles ... But who among us frolics in the sea-foam and daydreams of blood-red, human-size Humboldt squid? Sarah McAnulty, that's who.

"I think the thing that has hooked me [about squid] is that you can look at an animal that is behaviorally supercomplex, but diverged from us so, so, so long ago in evolutionary time," says McAnulty, a squid biologist and executive director of an educational nonprofit called Skype a Scientist.

Squid last shared a common ancestor with vertebrates (or animals with a backbone, like us) something like 600 million years ago. That's so long in the past, Earth days were hours shorter and Earth years were days longer, because our planet's position relative to the sun and moon has changed over time.

And yet, says McAnulty, despite hundreds of millions of years of evolution between us, squid are one of the very few invertebrates that make eye contact with humans. “They will swim over to you and be like, ‘Hey, what are you doing?’” she says.

On top of that, squid can glow, change colors, and even leap out of the water and soar through the air using their mantle flaps as gliders (the Humboldt squid is also known as the jumbo flying squid). The smallest squid species make up the basis of entire food webs, while the largest fill the bellies of sperm whales. “Squid are like swimming protein bars, because they don’t have bones to deal with,” says McAnulty. “So they’re very, very easy to eat, and so almost anybody who has the opportunity will eat squid if they’re the right size.”

And when you’re talking about Humboldts, well ... “Sometimes they eat *each other,*” she says with an air of respect.

Diablos Rojos

Named for the Humboldt Current off the Pacific Coast of South America where the species was first described, Humboldt squid are one of very few squid species to have gained a reputation for being dangerous to humans.

“There are certainly fisherman stories of people getting pulled overboard and eaten by Humboldt squid,” says McAnulty. The good news is there are no confirmed accounts of this actually happening. But it’s not entirely outlandish to consider.

For starters, Humboldt squid look a bit like a torpedo dragging 10 garden hoses, each of which is lined in barbed wire. Technically, eight of these extremities are called arms, says McAnulty, and each of those is covered in up to 200 suction cups ringed with little teeth (other squid in the Humboldt’s subfamily have fewer than 35 suckers per arm). The two remaining extremities are known as tentacles, and they’re easy to identify because the tentacles are longer than the arms and end in a paddle-like club, itself covered in teeth-bedazzled suckers. The clubs also feature literal hooks, which the squid sinks into its prey before reeling in and devouring the victim with a sharp, parrotlike beak and a radula, or tongue covered in tons and tons of *even more* tiny hooks.

If you know a good therapist, send that person the Humboldt’s way, because these things are not built for letting go.

Humboldt squid vary in size each year, and adults often have mantles (think torsos) that are no longer than a loaf of bread. But when lots of food is available and the temperature is cozy, the biggest Humboldts can measure more than eight feet in length.

Now, during the day, when you're sipping piña coladas on the beach, Humboldt squid hunker down near the twilight zone, or the depth of the ocean at which sunlight fizzles out. It's only at night that they rise, like clouds of red-and-white rockets, out of the abyss and into the realm where humans have a chance of spotting them. (Don't worry. This happens on the open ocean, not by the shore.) The squid rise because their prey are moving upward in the water column, too. In fact, so many creatures large and small perform this nightly ascension, the phenomenon has its own name—diel vertical migration—and it's considered the largest coordinated movement of animals on Earth. Of course, most of us never see the spectacle because we're at home in our beds. But fisherfolk know it well.

In Mexico, nighttime fishing brings in more Humboldt squid than almost anywhere else on Earth. In some years, the annual haul weighs more than 10,000 tons, most of which gets turned into fish bait, steaks, or canned squid products. Much of the catch belongs to artisanal fishers, who use baited handlines and lights to lure Humboldts to the surface. It works because the squid's favorite food, lantern fish, also light up by way of bioluminescence. So when the Humboldts see light, it might as well be the neon sign of a drive-through restaurant. And they come swimming in groups that can be 1,200 animals strong.

"I remember being in the Sea of Cortez during the day and looking at the sonar at about 900 feet, and it was solid red for miles," says Brian Skerry, a National Geographic photojournalist and one of the few people to document nighttime squid fishing from below the water's surface. "And that was all Humboldt squid."

Scientists have also noted that the jumbos sometimes seem to be working together as they hunt, each flashing secret codes with their skin that seem to announce which fish they're going to attack next, perhaps as a way to save energy or maybe avoid friendly fire. After all, as McAnulty alluded to earlier, Humboldts have been shown to dabble in cannibalism.

With special color-changing cells called chromatophores in their skin, the squid can turn ghostly white in an instant. Or half white and half crimson. With just two colors, they can create nearly 30 different patterns, from racing stripes and spots to strange, mottled mixtures. Combined with body posture and context, such as whether the squid is swimming with other squid or in the process of feeding, scientists believe the animals use these color flashes to communicate with each other, warn off predators, and disorient prey as they strike.

Some of these patterns occur in the same order so frequently that it reminded the scientists of the way we arrange words in sentences. Best of all, when the squid are down in the twilight zone, their entire bodies can glow, which backlights the patterns and makes them visible where the sun's light can't reach.

Now, let's remember that those who want to catch Humboldts must go in search of them at night. In the middle of the ocean. And also that, due to its reputation and coloration, the animals are known in Mexico as *diablos rojos,* or red devils. Add it all up and you've got, well, if not a recipe for disaster, then at least all the fixings for some campfire-worthy kraken stories.

"I wouldn't swim with a giant squid, but in the same way that I wouldn't put myself near any big, unpredictable predator," says McAnulty. "But you don't need to be afraid of deep sea squid, because they don't live near you."

Of course, fishers aren't the only ones who intentionally venture into the red devil's domain. Skerry says he's spent his entire life exploring the oceans, documenting marine life, and trying to communicate that magnificence back to the public in a way that creates respect rather than fear. "But you know, the Humboldt squid is the one animal I would say really fits the description of a sea monster," he says.

In fact, Skerry remembers the moment he told one of his friends, esteemed squid biologist Roger Hanlon, that he planned to dive with Humboldts. "And he said, 'Brian, whatever you do, do not dive with them. One of those squid could swim away with you and me both.'" Fortunately, Skerry's shoots with Humboldts left him unscathed. But other photographers have suffered serious injuries.

Is now a good time to tell you that, in just the past few decades, Humboldt squid have started showing up in places no one had ever seen them before? Though the predators are typically found from Baja California all the way down to Chile, they were spotted in 1997 hanging out in Monterey Bay, California. And in the late 2000s, Humboldts became a regular presence in Washington's Puget Sound. Now, the red devils are sometimes seen as far north as Alaska. Scientists are still trying to disentangle all the reasons why. Climate change and warming oceans may expand the comfort range of some species, while shrinking it elsewhere, says McAnulty. At the same time, overfishing of common squid predators such as tuna can create a more accommodating environment for them. "Many squid species are moving north," say McAnulty.

Again, you and I have nothing to worry about from Humboldt squid. But there's something rather magnificent about knowing that every night, an army of hundred-pound red devils rises up out of the depths, flashing their secret patterns and devouring everything they can sink their hooks into.

THE MYSTERIOUS MIGRATOR

MONARCH BUTTERFLY

TYPE: Insect **SPECIES:** *Danaus plexippus* **STATUS:** Vulnerable **RANGE:** Primarily from southern Canada to central Mexico **SIZE:** 4-inch wingspan; weighs up to 0.02 ounce **LIFESPAN:** Up to 8 months

---

PLACE TWO STICKY NOTES in an open palm, and you'll have a pretty good idea of what it's like to hold one of the world's most recognizable insects—the migratory monarch butterfly. These beauties can have orange-and-black wingspans that stretch up to four inches, but they only weigh about as much as a short stack of fleas.

Once each year, millions of monarchs leave their quiet little lives in the United States and Canada and take to the sky. For around two months, these insects—which look as if they'd have a hard time crossing the road—fly around 50 to 100 miles a day until they reach the mountains in central Mexico's Volcanic Axis. There, they will cluster so densely on oyamel fir trees that it looks as if the trees have blossomed with orange flowers. And while each monarch may weigh next to nothing on its own, together they can achieve enough mass to make the giants quake.

"They actually make the branches of the firs bend," says Jorge Rickards, director general of World Wildlife Fund México.

Rickards has been traveling to witness this wildlife spectacle since he was a young boy. Back then, the monarchs were well-known to the people of Mexico, where some believe the winged ones are more than insects.

"It goes back hundreds of years," says Rickards. "It is seen as the souls of the dead people coming back to visit the living."

And no wonder. The monarchs start arriving in the mountains outside of Mexico City in early November, and Día de los Muertos (Day of the Dead) is celebrated between November 1st and 2nd. It's also handy that the butterflies are similar in color to orange marigolds, which are known as the flowers of the dead and thought to help guide souls to the altars laid out for deceased family members.

What's wild is that when Rickards was a kid in the 1970s, people in Mexico didn't know where the monarchs drifted off to when they left the highlands in the spring. At the same time, scientists in the United States and Canada couldn't say for sure where the monarchs had been all winter. Everyone on the continent had a piece of the puzzle, but it wouldn't be until 1975 that those pieces came together. Fifty years later, we have the great privilege of knowing where the monarchs go. And how they get there is nothing short of mind-boggling.

A Journey From Goo to Greatness

When an Arctic tern or a caribou migrates, it leaves an area where the weather is turning for the worse or the food is running out and heads for a place where conditions are better. Then, after some time, it turns around and goes back again.

But a monarch's migration plays out over the course of multiple generations each year, which means no single butterfly ever sees the whole thing. To understand how it works, let's begin 10,000 feet up the side of a dormant volcano in the Monarch Butterfly Biosphere Reserve in Mexico, a UNESCO World Heritage Site.

As ectotherms, butterflies rely on external temperatures to get their bodies primed and ready for motion. And when the sun creeps over the horizon and starts to warm the insects in their clusters, one by one the monarchs will begin to take flight until millions of orange-and-black flappers fill the air.

"It's like orange confetti flying all over the place," says Rickards. "You can actually hear them flapping."

But marvelous though it may be, the spectacle can't last forever. And by March, the monarchs have moved on, journeying back across the border to the north to seek out nectar-producing flowers and the plants known as milkweed across the American South.

The migrating monarchs' fate is inextricably linked to this family of plants—and the loss of milkweed across the American heartland has been pegged as one of the largest reasons for monarch decline in recent decades (other threats include logging near the Mexican wintering areas, climate change, and pesticide use).

Milkweed can be noxious to grazers like cattle and even cause contact dermatitis in some people, but the plant acts like a nursery for the monarchs' babies. So when a monarch mama finds the appropriate milkweed plant, she will float down and lay a single, pencil-point-size egg on a leaf. In just a few days, that egg hatches into a tiny, striped caterpillar that has just one job: munch as many milkweed leaves as it can. Indeed, these plants are both food and ammunition, because the caterpillars actually sequester the milkweed's toxins and use them to make themselves poisonous—a trait that stays with them through adulthood.

In a few days more, the caterpillar will have gorged so hard that it is ready to burst out of its skin. This is known as molting, and each time a larval insect performs the task, it levels up to a new stage of instar. Monarch larvae pass through five instar stages, growing up to 100 times in size. When all is said and done, they look like brilliant black-white-and-yellow-striped sausages.

Next, the caterpillars will spin a silk pad on the underside of a leaf and anchor themselves to it before hanging upside down and shedding their skin one final time. Locked in place and all scrunched up, their body hardens into a bright green alien pod known as a chrysalis. (Note: The other way to do this is to spin silk all around yourself, which is called a cocoon.)

And now, things get super-duper weird.

You know from elementary school science class that the caterpillar is turning into a butterfly. But, uh, those textbooks left out some details about what's known as the pupal stage. Starting with the fact that, inside the chrysalis, the caterpillar dissolves into a sack of living goo.

Enzymes inside the caterpillar digest the insect from the inside out, leaving behind only the imaginal discs, which are sort of like foundation sites for the structures that come next.

Special cells in the monarch slime flock to the imaginal discs and start to go gangbusters, dividing rapidly and building adult butterfly parts such wings, antennae, legs, and eyes.

Open up a chrysalis in the middle of the 10 to 14 days it takes to complete its metamorphosis and you will find neither caterpillar nor butterfly, but rather a sort of mortal sludge.

Can the monarch comprehend what's happening to its body? Likely not, given that its eyeballs and other sensory organs have been broken down and cannibalized to build its final form. But at the same time, this sac of goo is kind of, sort of ... sentient. At least in the sense that it is capable of transporting memories from the caterpillar to the butterfly.

How could we possibly know such a thing? Well, because scientists have taught tobacco hornworms to dislike a smell they've never smelled before (by exposing them to a mild electric shock while in the presence of it) and found that the animals actually remembered and recoiled from that same novel smell after they had turned into adult moths. This despite the fact that their brains and nervous systems had been dissolved and reorganized in the process!

Once the rejiggering is complete, the monarch will burst out of its chrysalis and inflate its wings by pumping fluid into the wing veins. In time, the wings harden, and the insect is ready to feed on the nectar found in flowers, and generally continue the migration northward, where the insects once again mate and lay eggs.

These eggs also hatch and undergo the miracle of metamorphosis, as monarchs do, before heading north again. And their offspring do the same. Until, by the end of summer, millions of monarchs will have returned to the American Midwest, Northeast, and southern Canada. (By the way, there are also monarchs west of the Rocky Mountains that perform a very similar migration, but one that ends in California rather than Mexico.)

But then, in the fall, something else bizarre happens.

Unlike all the four to five generations of northward-flying butterflies that came before, the monarchs that hatch now form a single "super generation." And unlike all the silliness surrounding the supposedly inherent qualities of human generations—Boomers! Gen X! Millennials! Zoomers!—these butterflies are actually fundamentally different from the butterflies that preceded them.

How so? Well, the super generation is also known as the Methuselah generation, after the guy in the Bible who is supposed to have lived to the age of 969. And the reason for this is simple: Super generation monarchs live eight times longer than other monarchs, which might only live for two to six weeks.

The supers also travel 10 times farther as they skitter back toward the sacred firs of Sierra Chincua—a place they've never been before and will never go again. A place where the whole spectacular cycle begins anew.

THE SEE-THROUGH HARPOONER

MOON JELLY

TYPE: Cnidarian **SPECIES:** *Aurelia aurita* **STATUS:** Not Evaluated **RANGE:** Most coasts of North America; also in waters off Asia **SIZE:** Up to 2 feet in diameter; weighs up to 1.6 ounces **LIFESPAN:** Up to 1 year in the medusa stage

---

ON PRETTY MUCH EVERY COAST of North America, you can find dinner plate–size creatures that are, statistically speaking, living water. They have no bones, no brain, and gonads that resemble four-leaf-clover tattoos, and yet, these otherworldly invertebrates boop through the ocean in mesmerizing, unremitting motion, as if marching toward some destiny known only to them.

"They can move. They can make decisions. They can swim against the current a little bit," says Rebecca Helm, a marine biologist at Georgetown University's Earth Commons Institute. "But they can't really control where they're going in a broader sense, like a whale might. That's why we call them drifters."

Now, you might be inclined to think of drifters as lesser than. After all, when you see a flying fish or a dolphin careen through the waves like a bullet fired from a gun, a creature that can merely bob about looks a bit inefficient or even ridiculous by comparison. But jellyfish know what they're doing.

"In essence, jellyfish aren't really trying to move. They're trying to move water *to them,*" explains Helm, who is also a research associate at the Smithsonian's National Museum of Natural History. Every shiver, every undulation of their bell, every flourish of their tentacles is an attempt to create vortices of water that usher tiny crustaceans, fish eggs, larvae, and other microscopic life-forms ever closer to their strike zone. Once prey is in range, the moon jelly's tentacles sting the prey and draw it to the mouth on the underside of the bell, where digestive cells go to work. (Silly aside—the jellyfish's mouth is also its anus. Waste comes out the same orifice it went in, in other words.)

Even still, Helm says, "They're supergood at what they do. We're judging them by the wrong metrics."

So let's talk about something everyone must respect—the venom.

Unlike rattlesnakes or yellowjackets, which have fangs and stingers you can see, a jellyfish's venom-delivery devices are made up of microscopic stinging cells known as nematocysts. Helm likens the structures to tiny mason jars stuffed with mini-harpoons and primed with hairlike triggers on top.

"When you brush against the tentacle, you trip that trigger, and the harpoon shoots out and injects venom into your skin," she says. "And each tentacle is covered in thousands of cells."

Now, some jellies' venom, such as that of the box jellyfish native to Australia and the Indo-Pacific, can knock a human dead. But moon jelly venom isn't lethal so much as it is annoying. On the cellular level, the stuff can destroy red blood cells and break down proteins, which sounds scary, but really it just leads to an intense stinging sensation that can also burn, itch, and welt. Not something you want to experience, obviously, but it's probably worse for the jellyfish.

"When I've had jellyfish in captivity and I've been stung—on accident, because I wasn't being careful—one of the amazing things I noticed was that the jellyfish didn't do very well after they stung me," says Helm. "They got smaller. They shrunk."

Why? Because the only way a jellyfish can eat is by stinging stuff with its tentacles and then reeling it into the mouth. If it wastes those stings on you, it has fewer stings left for lunch.

"They definitely never, ever, ever want to sting a human," says Helm. "So anytime you have an encounter with a jellyfish and you get stung, I can assure you that it was unpleasant for everybody involved. Including the jellyfish."

While we don't tend to think of them as such, jellies are critical components of the ocean food web. For starters, some animals eat nothing but jellyfish, such as leatherback sea turtles. So no jellies, no leatherbacks. But jellyfish are important predators themselves, gliding through the oceans and vacuuming up innumerable bits of life. The truth is, so much of this happens outside of our view that we don't even know how to quantify what would happen if the jellies went away—but it probably wouldn't be good.

"Jellyfish are like mobile nursery grounds for a lot of different species of fish," Helm says, referring to the way the jellyfish's tentacles protect such creatures from predators. "All sorts of animals depend on jellyfish."

Of course, it's nonetheless possible to suffer a jellyfish sting, which nobody wants. But if you've been stung, don't pee on it, as folk wisdom might advise. For one thing, it's gross. For another, doing so can actually make the sting worse, because urine can cause the nematocysts that are still embedded in your skin to start firing again and releasing more venom. A better approach would be to use a credit card or a handful of sand to scrape across the site of the sting. "Your goal is to remove any extra bits of tentacle or jelly tissue," says Helm. Next, apply vinegar, meat tenderizer, or a product called Sting No More. "These will break up any microscopic stinging cells left behind," she says. Finally, a hot pack applied to the area helps to break down venom that's already inside your body.

How Jellies Are Like Trees

When most of us picture jellyfish, we think of globular bodies with strings of tentacles among pulsating swarms drifting through the ocean. But this setup only describes a third of the animal's overall existence.

Jellies undergo metamorphosis, like butterflies and frogs. And the large, gelatinous things we call jellyfish are only one phase in the greater cycle. In fact, this well-known phase is the only one scientists refer to as jellyfish. If you want to talk about the whole animal, "jelly" would be a better term. More specifically, you could also call this phase the medusa or sexually reproductive form.

As adults (medusae), jellyfish release either sperm or eggs into the water (yes, there are boy jellies and girl jellies). When an egg and sperm meet and fuse together, they create what's known as a planula larva, which Helm says looks like a fuzzy pill swimming through the water column. No bigger than a poppy seed, these larvae drift down to the ocean floor in search of a nice safe place to settle down and transform into the next life stage, known as the polyp.

"Polyps are about the size of breadcrumbs. They're milky white in color, and their body is shaped like a flower," says Helm. "A cup-shaped body with a ring of tentacles around a central mouth." Moon jellies can spend years sitting on the seafloor as polyps, but when conditions are right—a process that seems to involve changes in water temperature triggering protein production and then hormone release—the polyp prepares to undergo yet another metamorphosis, called strobilation. And this is where things get wacky.

When a tadpole turns into a frog, it makes one frog. When a caterpillar turns into a butterfly, it makes one butterfly. But when a moon jelly polyp turns into a strobila, it becomes a launchpad for a whole mess of new baby jellies. Just imagine a stack of pancakes where the top pancake keeps peeling off and swimming away (once the little guys swim away, they are known as ephyra, at least until they mature into medusae). Meanwhile, the last pancake in the stack sticks around, anchored to the seafloor where it can turn back into a polyp and start the strobilation process over again when conditions are right.

"One polyp can make tens or even hundreds of jellyfish," says Helm. And each of them is a clone of the original.

By the way, this whole strobilation thing qualifies as what scientists call asexual reproduction. And earlier, when the adult jellyfish released sperm and eggs? That was sexual reproduction. Which means that, over the course of their lives, moon jellies do it both ways.

"In a sense, jellies are a lot more similar to what we're used to seeing in things like plants," says Helm. In other words, you've got one tree that blooms into hundreds or thousands of flowers, each of which gets pollinated, then transforms into seeds (or fruit with seeds inside), all of which fall off and go on to create new trees. Meanwhile, the original tree stands ready to repeat the process again next year. That sounds a heckuva lot more like jellyfish reproduction than it does mammal reproduction, at least.

One species, known as the immortal jellyfish *(Turritopsis dohrnii),* takes all of this strangeness a step farther. When adults (or medusae) die, their cells reassemble into polyps, which can then settle and transform into strobila. Which means that under the right conditions, an immortal jellyfish never dies, it simply shuffles on to another life stage.

Moon jellies have been observed performing a similar feat in captivity, but whether or how much moon jellies cheat death out in the wild, scientists can't yet say. Clearly though, the animals are already living many lives at once as they cycle and serve up clones of themselves.

Taken with the revelation that jellies don't swim to travel but rather swim to eat, it has me thinking about that old Ralph Waldo Emerson quote: "Life is a journey, not a destination." Or maybe that was Aerosmith. Either way, the jelly journey is a wild ride.

RAVENOUS ROAMING PINCUSHION

PURPLE SEA URCHIN

TYPE: Echinoderm **SPECIES:** *Strongylocentrotus purpuratus* **STATUS:** Not Evaluated
RANGE: West Coast from Canada to Mexico **SIZE:** Up to 3 inches in diameter; weighs up to 1.9 ounces **LIFESPAN:** Up to 50 years

---

JUST BELOW THE WAVES on North America's Pacific Coast, an army of purple pincushions is on the march. And while we might need a time-lapse camera to see it move, this creeping, crawling mass of spines has the power to remake worlds.

The purple sea urchin is an underwater invertebrate native to the intertidal zone from British Columbia to Baja California. There's also an Atlantic purple sea urchin that can be found from Maine to the Yucatan Peninsula, but it's technically a different species known as *Arbacia punctulat.* In fact, many other kinds of urchins have "purple" in their common names, but this one has the color built into its scientific name, *Strongylocentrotus purpuratus,* though it also comes in a variety of reds and reddish-purples.

They don't look much alike, but urchins are most closely related to sea cucumbers and sea stars. All these ocean dwellers hail from an extremely primordial branch of life known as the phylum Echinodermata. Though this sounds like some sort of religious affliction, Echinodermata roughly translates to "hedgehog skin" in Greek and refers to the spiky protrusions common in each of these critters.

“The phylum itself dates back to the Paleozoic,” says Christopher Mah, an invertebrate zoologist at the Smithsonian’s National Museum of Natural History. In fact, the earliest known echinoderms lived around 450 million years ago, which means they showed up long before life had the gumption to try things out on land.

Within this motley crew of hedgehog-skins, urchins leaned hardest into the moniker and came out the other end of evolution with weaponized skeletons meant to keep predators at a spine’s length. Notice I said skeleton, not shell. Urchins don’t have shells, says Mah, because shells are external structures. Experts like Mah call the internal structure a “test.” On the outside, the critters sport a layer of living tissue, or epidermis, just like you and me.

The creatures we call sand dollars, as a matter of fact, offer a great view of the test. Sand dollars are actually another kind of urchin, one that burrows in the muck. Live sand dollars are coated in living tissue and pointy structures like sea urchins, but by the time we see them, the ocean has cleaned them of all that decaying material so nothing but the flattened test remains.

For “regular urchins”—an actual scientific distinction from irregular urchins, such as sand dollars—the outer coating of skin is important because an urchin’s epidermis is also its sensory system. Urchins lack eyeballs, but their spines can sense light, and they can even use differing information across that spinal network—it’s bright over here, dark over there—to orient themselves and decide which direction to move.

While lots of people think that urchins are stuck in one spot, like a barnacle or giant clam, urchins are actually capable underwater rovers. They just happen to move at a pace similar to what my toddler calls “tortoise mode” when he’s being difficult.

“All echinoderms depend on tube feet and what’s called a water-vascular system,” says Mah. You can think of it like hydraulic tubes snaking throughout their bodies. Each tentacle ends in a little adhesive-covered disc, says Mah, which allows urchins to grapple across the seafloor, rock-climb ledges, and even scale stalks of underwater vegetation. Tentacles can also pass food collected on the urchin’s back all the way down to its mouth.

While these tube feet tentacles are often hidden on the underside of sea stars, they're especially noticeable in urchins because they radiate outward from the test in every direction. That is, if the urchin feels comfortable and safe. When threatened, they can draw all the tentacles back within the perimeter of the spines.

While urchins don't technically have a front or back, they do have a top and a bottom. And if you were to flip one over, you'd find an alien-like orifice made up of still more tentacles, rings of flesh, and a five-part beak. This is the purple sea urchin's mouth, and each of those five parts is composed of teeth-like structures that can move independently of the others. For the urchin to take a bite, all five come closed at the center, with a nearly imperceptible clockwise flourish, like a star's points collapsing inward.

Add it all up and this glorified snippety-snip device is known as an Aristotle's lantern, because the ancient philosopher thought it looked liked the five-paned lanterns popular in his day. And though urchins can't bite people, when you get millions of these sea hedgehogs in the same place, the combined wrath of their Aristotle's lanterns can fell entire underwater forests and bring vast ecosystems to their knees.

The Urchin and the Damage Done

The story for some species in this book is one of catastrophic decline, but other species prove the adage that you can have too much of a good thing.

Purple sea urchins are basically aquatic lawnmowers. They're herbivores, and they make a living by gnawing and nibbling underwater vegetation such as plantlike algae and towering stands of kelp. Urchin herbivory is important because without someone coming by to trim the garden, undersea flora can become overgrown and unusable for creatures that rely upon it. Too much algae, and coral can't colonize new areas, for instance.

On the flip side, if you have too many grazers, they can devour every last scrap of greenage in an area, turning what was once a lush, underwater forest into a wasteland. And urchins are so dang good at producing such areas, scientists came up with a special name to describe the carnage: urchin barrens.

When an ecosystem is in balance, herbivores eat the plants, predators eat the herbivores, and still more predators eat those predators. And despite being spikier than a bed of nails, urchins do have natural predators that can keep their numbers in check.

Sea otters break off the spines, bite through the test, and then lick out an urchin's insides like you would an oyster. And some otters on the West Coast eat so many urchins, their teeth, skulls, and bones actually turn purple from all the pigment. This harmless phenomenon is known as echinochrome staining. (By the way, people who step on urchins often receive a similar souvenir—a tiny pinprick tattoo left behind everywhere the spines punched through flesh.) There's also a three-foot-wide, predatory sea star known as the sunflower sea star that uses its two dozen arms to flip a purple sea urchin over, lower its stomach into the urchin's mouth, and then digest it from the inside out. "They eat them on the half-shell," jokes Mah.

Unfortunately, sea otters were hunted for their fur to the point of near-extinction in the beginning of the 20th century. And sunflower sea star populations crashed between 2013 and 2017, when a devastating bout of sea star wasting syndrome marched north along the Pacific Coast, causing around 20 sea star species to melt into piles of goo. And these factors, combined with a few others—a giant influx of warm water, an El Niño–related heat wave—created conditions that have allowed urchin numbers to skyrocket. And those urchins have done what urchins do. They feasted.

As a result, satellite imagery reveals places in Northern California where kelp forests have declined by as much as 95 percent. And while the factors may differ depending on where you go, there are now plenty of places on the West Coast where urchin barrens have replaced kelp forests.

The good news? Thanks to conservation efforts, otter populations are bouncing back. And in areas of California's Central Coast, where otters now roam, a different scenario has played out. Even though these waters were hit by waves of warmth and the urchin armies that followed, the otters seem to have had a mitigating effect on the damage done. There are still barrens, yes, but also large sections of healthy kelp forest—a quilt-like pattern of varying habitat scientists refer to as a mosaic.

Interestingly, in 2022, scientists discovered that a single-celled parasite known as a ciliate had been wreaking havoc on different species of urchins living off the coast of Florida and in other parts of the Caribbean. In 2023, the same culprit was implicated in urchin die-offs in Oman on the Arabian Peninsula. And while too many urchins can be a problem for kelp forests, with too few, algae (upon which urchins graze) can go bananas, in turn smothering and suppressing the growth of coral reefs. In Hawaii, scientists have actually released one million captive-bred native urchins onto the reefs as a way to fight an invasive species of algae. There, urchins are the heroes of the story.

All of which is a good reminder that no animal is intrinsically *bad*—not even a madcap mass of tentacles and spines.

THE WORLD WANDERERS

DRAGONFLIES

TYPE: Insect **INFRAORDER:** Anisoptera **STATUS:** Varies among family, from Least Concern to Endangered **RANGE:** Global, including most of North America **SIZE:** Wingspan up to 6 inches **LIFESPAN:** Up to 7 years

---

WITH MORE THAN 300 species in North America, dragonflies are well-known for being brightly colored and fast-flying aerial maniacs. Each species differs in details, but most follow a pattern. Their wings are see-through, their body is shaped like an ice pick, and their legs are spindly as all get-out. But the main attraction on any dragonfly is that faceful of eyeballs.

"There's not really much else in their head, other than their eyes," says Jessica Ware, curator of invertebrate zoology at the American Museum of Natural History. "And that is presumably because they've been selected to be highly visual predators."

It is not an exaggeration to say dragonflies are flying killing machines. And I mean that in the nicest possible way, of course. House cats are bird-killing machines. Ladybugs are aphid-killing machines. Perspective colors the way you see the world. And actually, dragonflies have more perspective than most.

Proteins called opsins allow us to see color, and human eyeballs come equipped with three opsins that make the red, green, and blue regions of the light spectrum visible. But dragonfly peepers have between 15 to 33 opsins, which hints at their ability to perceive far more than we can even imagine. Interestingly, some of these opsins turn on when the insects are larvae living underwater, and others while they're adults soaring through the sky.

Dragonflies lay their eggs in water—everything from puddles to ponds and streams. (This practice is also how the skimmer, a distinct group within the dragonfly world, got its name—because the females skim across the water's surface, dipping their abdomens in every now and again to deposit eggs on the fly.) Young dragonflies hatch into those environments and spend their childhoods there, lurking in the shadows and snatching animals that wander too close.

Look closely at a dragonfly larva and you will find mouthparts that are kind of like a hydraulic bear trap. When prey is near, the larva pumps hemolymph (insect blood) into those mouthparts, which makes them reach out and latch on incredibly quickly, says Ware. Then when that pressure is released, the array of snatchy mouth-arms hauls the unfortunate soul back into the dragonfly larva's mouth where it can be torn to shreds. Even as babies, dragonfly larvae can take down comparatively huge creatures, such as fish and tadpoles. They are also super helpful to humans as voracious predators of mosquito larvae. "They also eat each other," says Ware, an expert in the order known as Odonata (which includes both dragonflies and their close cousins, damselflies).

Of course, to be able to do all this, the dragonfly larvae need to first be able to detect their prey. And as anyone who has ever opened their eyes underwater without goggles knows, seeing through water is fundamentally different from seeing through air. Our eyes clearly aren't evolved to do both. But dragonfly eyes do so swimmingly, thanks in part to this array of opsins.

Interestingly, those opsins are also expressed differently in the tops and bottoms of adult dragonfly eyes—which means they can see certain wavelengths above them, where the sun is brightest, and others below, where light is dimmer and reflects off leaves, grass, and water. We can imagine dragonfly vision kind of like wearing bifocal glasses. Or rather, 15,000 pairs of bifocals, since these critters also boast up to 30,000 ommatidia, or curved, lenslike structures that give dragonflies (and bees and cockroaches and flies) mosaic vision. In other words: "They're able to see in a lot of different directions at many different angles," says Ware.

In addition to sight power, dragonflies also possess remarkable flight power. The insects' four wings can move independently of each other, which allows them both to hover and move in any direction at speeds up to 35 miles an hour. In fact, when it comes to flight in general, dragonflies have been doing it longer than any other creatures on Earth. "We think that flight first arose in insects around 400 million years ago," says Ware. "And probably, whatever that first thing to fly was, it was something like a dragonfly. It was either an actual dragonfly or a proto-dragonfly." So it's no surprise that modern-day dragonflies are absolute midair menaces—their ancestors started honing their flight maneuvers many millions of years before the Rocky Mountains rose out of the sea.

If dragonfly larvae are sit-and-wait predators, then the adults are heat-seeking missiles. "It's called interception predation," says Ware. Instead of heading to where the prey is, dragonflies zip straight to where their lunch *will* be. And this strategy allows them to capture mosquitoes, blackflies, horseflies, butterflies, aphids, and even other dragonflies with up to a 97 percent success rate. Just to give you a sense of how staggering that kill rate is, note that cheetahs succeed in 58 percent of their hunts, leopards in 38 percent, and wolves in just 14 percent.

Perhaps the most impressive of all? While all of those mammalian predators would have to undergo long periods of rest after a hunt to recoup the energy lost from the chase, dragonflies don't even stop flying. They gobble their prey on the wing and then shift their focus to the next target. "You can see dragonflies sometimes in the late afternoon just going backward and forward, backward and forward," says Ware. "We call it hawking behavior, and they are basically feeding on aerial plankton, all the flotsam and jetsam that's in the air."

Only when the sun sets will these fighter pilots power down on a reed or tree trunk. And sometimes, not even then. Because one kind of dragonfly is unlike any other on Earth: the globe skimmer.

Globe Skimmers Get Around

Over the course of her career, Ware has had the good fortune to travel all over the world to study and collect insects. But no matter where she goes, there are always globe skimmer dragonflies. Globe skimmer dragonflies have a wingspan of slightly more than three inches and a gorgeous orange-and-yellow iridescence that makes them look like an oil slick come to life.

"Whether it was Canada, or Australia, or Namibia, or Costa Rica, when I went places, I saw [globe skimmers]," says Ware. "And I thought, 'This is so interesting. This thing is everywhere!'"

The globe skimmer's cosmopolitan distribution is well documented in scientific literature. Records from sailors dating back to the 1800s mention sky-darkening swarms of the insects hundreds of miles out at sea—a place most insects should not be. "Saltwater is death for dragonflies and damselflies," says Ware.

Also curious was the fact that globe skimmers showed up on extremely far-flung islands, such as the sub-Antarctic Amsterdam Island in the Indian Ocean. Mountaineers even spotted globe skimmers while climbing the Himalaya, at elevations of nearly 20,000 feet. No wonder globe skimmers are known as the wandering glider.

To figure out how these dragonflies could seemingly be everywhere all at once, Ware started lighting their wings on fire. Don't worry—this wasn't an interrogation method but rather a method for measuring their chemical composition.

Conveniently, the hydrogen isotopes found in a dragonfly's wings are acquired from the waters of their youth, which means each insect is carrying the equivalent of a homing beacon. This is because hydrogen varies in its atomic weight depending on where you are. "You can see, does that hydrogen have the weight of Vancouver water? Or does it have the hydrogen weight of Sierra Leone water?" says Ware. "We found that almost all of the individuals that we sampled were from other continents than we'd caught them on as adults."

What's more, when Ware and her team looked at the genetics of globe skimmers sampled in 29 countries on six continents, they discovered that all the creatures belonged to the same species, rather than being just a bunch of red and yellow dragonflies that looked the same. They could also show that globe skimmer genes were flowing virtually every which way across Earth, from the Eastern Hemisphere to the Western, from the Northern Hemisphere to the Southern. Globe skimmers got around like maybe no other animal. According to Ware, this makes globe skimmers one giant pool of dragonfly-dom, or what scientists call a single global panmictic population.

Interestingly, all of this tracks with another study that proved globe skimmers fly from northern India to the Maldives and then on to east Africa before returning to India again—a multigenerational migration of about 8,700 miles, the longest known for any insect.

To live in the sky, globe skimmers probably feed constantly, replenishing fat stores that allow them to beat their wings nonstop for up to eight hours at a time. When tired or resting, the insects seem to be able to drift vast distances via high-altitude winds, and this allows them to spend months, maybe even years, aloft. No one knows how long a globe skimmer lives, after all—only that they breed in temporary puddles left behind by heavy rains. While other dragonfly species spend up to 12 months or more as larvae, globe skimmers bebop through the life stage in just four to six weeks. In part, globe skimmer development is accelerated because they're constantly in danger of their pools drying up. But it's also almost as if the sky calls to them, and they simply cannot wait to disappear into it.

THE HAIRY-BELLIED BUDDHAS

TARANTULAS

TYPE: Arachnid **SPECIES:** More than 100 species in North America in the family *Theraphosidae*
STATUS: Varies among family, from Least Concern to Critically Endangered
RANGE: West of the Mississippi River to California; all of Mexico **SIZE:** From 1 to 7 inches in diameter; weighs up to 6 ounces **LIFESPAN:** Up to 28 years

---

EVERYTHING YOU KNOW about the tarantula is wrong. Literally everything. When we see big spiders in movies and on TV, they're always on the move. They climb up walls, across floors, and up the arms, legs, and backs of whatever character is most likely to freak out about it. But those who study these animals or keep them as pets know the truth—tarantulas don't usually do much of anything.

In fact, tarantulas live out most of their days as mini hairy-bellied Buddhas sitting beneath the Bodhi tree. They seek not enlightenment but crickets, beetles, and the occasional lizard, which provide both nutrients and water—a precious resource in environments where tarantulas are common and liquid is rare. Even these earthly indulgences are few and far between though, and, after a good meal, some tarantulas may go a month or more without eating again.

"They're very good at conserving energy," says Brent Hendrixson, an arachnologist at Millsaps College in Jackson, Mississippi. "They don't generally venture far from their burrows unless starved or disturbed."

Find the biggest arachnophobe you know and show them a tarantula on the hunt. They won't be traumatized. They'll be bored to tears.

This is because North America's tarantulas are what's known as sit-and-wait predators. Rather than go out looking for trouble, they hunker down until dinner wanders within striking distance. Then, and only then, do they act, all at once and in a flurry of limbs, to grab their tiny victims and pierce their bodies with two ridiculously long fangs.

Of course, if you happen to awake from uneasy dreams one morning to find yourself transformed into an insect, then yes, you should be very afraid of tarantulas. Because evolution has gifted these creatures an entire apothecary's worth of secretions that both knock bugs out and turn the insides of any insect unlucky enough to become prey into goop.

First a tarantula's venom causes its prey's nervous system to go haywire, paralyzing the critter in the process. This is helpful, because tarantulas are surprisingly delicate creatures, and they cannot afford to wrestle with everything they eat. With the prey chemically subdued, a tarantula then secretes digestive enzymes that go to work breaking down and softening its tissues, like Drāno.

Why turn your steak into gruel? Well, spiders don't have teeth for chewing—those fangs look fierce, but they're the equivalent of hypodermic needles, not meat grinders. Nor do these creatures sport the sort of intestinal machinery needed for digesting things like chicken wings and tacos. But with venom doing the dirty work outside the spider's body, all these predators have to do is bide their time, then slurp up the liquefied meal like a protein shake.

Now, before you start getting that itchy feeling like something's crawling up your leg—you just checked, didn't you?—consider that there are no documented cases of humans dying from tarantula bites. Nor would a tarantula have a thing to gain by biting a person. That is, unless that person happens to be harassing the spider.

"I have been handling tarantulas for nearly 40 years, and I have never been bitten," says Hendrixson.

By the way, Hendrixson got his first pet tarantula at age four. So if an excited, impulsive toddler can manage to avoid goading a giant spider into acts of violence, then the rest of us should have no problem staying bite free—even if we're around the spiders all the time.

The Great Tarantula Migration

Now that you have allowed your preconceptions about tarantulas to fall away, it's time to talk about an exception to the picture of solitude I just took some pains to paint above. Because every autumn on the high desert, brown tarantulas in the genus *Aphonopelma* emerge from their hovels and trek across the open plains in such huge, obvious numbers that the event is sometimes referred to as a massive tarantula migration. And tourists come from afar to witness it.

It's not actually a migration, though. It's a once-in-a-lifetime walkabout in the name of love.

Like I said, the vast majority of a tarantula's life is spent resting. They sit, they snatch, they suck. And every once in a while, they change their outfit. In a process officially known as molting, a spider must shed its skin, or exoskeleton, in order to grow. It's a natural part of how invertebrates increase in size, but the process looks kind of like a toddler trying to wiggle her way out of a onesie. In other words, it looks absolutely hilarious.

The actual event is deadly serious though, because while the spider is in the process of molting, it must remain mostly immobilized, making it an easy snack for predators such as snakes, lizards, coyotes, and birds. And that danger lingers for up to a week after molting, because the new coat of armor takes time to harden. Finally, the exoskeleton that must be shrugged off covers every nook and cranny on the spider's body, from the spinnerets it uses to make silk to the fangs and eyeballs up front, and there's always a chance that a piece will get stuck as the spider tries to bust out. This is more than a simple wardrobe malfunction, because within an hour, the old tissue will harden, potentially trapping the spider. For an arachnid, getting cinched in one's own skin can be a death sentence.

However, providing that our spiders find enough foods and survive their molts, things will go on like this for around a decade. It's a humble existence, and so fundamentally different from the pace of our own reality that, if you think about it long enough, you might find yourself pondering the nature of life itself.

But so far as we know, the spiders don't philosophize much. They're too busy gathering their will. Because after eight to 10 years of slog, the males undergo a final, glorious transformation.

"Right before they mature, the male tarantulas look very similar to the females.

They have the same build, with short, stocky legs," says Hendrixson. "I think that's a body form that's better adapted for living in the burrow." Compact. Powerful. Perfect for lunging at prey and hiding in a hole.

But something strange happens to the male spiders while they're undergoing that final molt. Their abdomens shrink down and streamline. At the same time, their legs become thinner and longer. And at the very tip of the front pair of legs, for the very first time in their lives, the males develop tiny spurs.

This is the equivalent of tarantula puberty. One day the male is a homebody with a short, squat, bodybuilder physique, and the next he's a long-legged marathon runner with a pair of grappling hooks for hands. And he only has one purpose left in life—to find a female and persuade her to mate.

Truth be told, we don't really know how male tarantulas find female tarantulas as they hide in their burrows. Especially since these spiders hunt by night and perform their mating reconnaissance by day. Perhaps the males are sniffing out pheromones, or chemical scents, given off by the females or the silk the females use to reinforce their homes. Maybe it's vibrations, hypothesizes Hendrixson. But whatever the case, one study that tracked male tarantulas with radio telemetry learned that the fellas can travel nearly a mile in their quest for a mate.

Once Tarantuleo meets Tarantuliet and encourages her to come outside—often, this seems to include a little free-form leg drumming—the couple play a game of spider patty-cake with their forelimbs. If everything checks out, the male then uses his spurs to lock onto the female, at which point he will transfer a small sperm packet. Once the deed is done, the male removes his hooks and runs away—not because he's a jerk, mind you, but because if he lingers, he can become a snack for the females. "It's actually kind of comical how fast they run," says Hendrixson.

If the male survives the encounter, he will wander off to try his luck with another female and another, until either one of the gals, a predator, or exhaustion snuffs out his spark. "Their coloration fades, and a lot of males will have lost legs. Who knows what sorts of stories they could tell," says Hendrixson. "I came across a male tarantula one time that had only two legs left. He was not doing very well."

It's a happier story for the females, though. After the mama-to-be uses the male's sperm to fertilize a clutch of eggs, she wraps the eggs in a silk cocoon. Then for the next few months, the female spider starts to act like a mama bear, guarding her young and even doting upon them, moving them up and down the burrow, perhaps to adjust their temperature and humidity. The attachment to her babies is so strong, Hendrixson says, that when they remove the egg sacs in the lab, the mother tarantulas appear to experience loss and hopelessness.

Back out in the wild, the spiderlings eventually decide it's time to move on, at which point they march out of the burrow in single file, like a column of ants. But they don't go far. Instead, they'll dig their own burrows near Mom's.

As far as we know, there's no more interaction between mother tarantula and her children. But there's also something kind of nice about the idea of all those tiny tarantulas growing up so close to home.

THE AERIAL HUNTERS

YELLOWJACKETS

TYPE: Insect **GENERA:** *Vespula* or *Dolichovespula* **STATUS:** Not Evaluated
RANGE: All of North America **SIZE:** Half an inch long **LIFESPAN:** 1 month to 1 year

---

YELLOWJACKETS HAVE A REPUTATION for being angry little wasps. For instance, that *bzz-bzz* they make? It's intentional. They want you to know they're there, not unlike the *rat-a-tat* of a rattlesnake. And of course, the threat is not idle. Female yellowjackets have barbed stingers that rapidly inject venom into their enemies, and, unlike honeybees, these picnic crashers can crack off a series of successive stings before you even know what's attacking.

For all of these reasons, lots of folks consider yellowjackets to be more aggressive and dangerous than their honeymaking cousins. In recent years, the general public has also seemed to soften on bees due to their importance as pollinators, as well as their role as victims of colony collapse disorder, or CCD, which is a phenomenon where all the workers in a colony suddenly go belly-up.

But uh, just one thing: Honeybees are from Europe. Which isn't to say colony collapse disorder is a good thing. It's just to say that we have tons and tons of pollinators that are native to North America that nobody gives a hoot about.

"Popular culture tells us that bees are important pollinators and wasps serve no purpose, so we are allowed to hate them," says Chris Alice Kratzer, an entomologist and author of *The Social Wasps of North America.* "But bees kill just as many people every year as wasps do."

Countless species of wasps are powerful pollinators in their own right. For instance, yellowjackets pollinate monarch-butterfly-sustaining milkweed, as well as asters, goldenrods, mints, and jewelweed. You can thank a yellowjacket the next time you eat a carrot, celery, sunflower seed, or flavor your food with oregano, basil, peppermint, lavender, fennel, dill, parsley, sage, rosemary, thyme, and any other foodstuffs you find at the Scarborough Fair. Because yellowjackets are important pollinators for all of the above.

And another thing. When the late, great entomologist Justin O. Schmidt compiled his list of insect stings and their intensities—officially known as the Schmidt sting pain index—he categorized yellowjacket stings as being worse than a fire ant but not as nasty as a bullet ant or tarantula hawk wasp. Ever the poet, Schmidt described the yellowjacket's sting as "instantaneous, hot, burning, complex pain that gets one's attention no matter what other thoughts were preoccupying the mind." But colorful language aside, Schmidt scored yellowjacket and honeybee stings equally—despite most people believing wasp stings to be worse.

So, what is a yellowjacket, exactly? Because that name gets bandied about for all sorts of black-and-yellow buzzy things.

Let's start with what yellowjackets are not. First and foremost, yellowjackets are not bees, though they are related to them. Bees reside in the superfamily Apoidea, tend to be vegetarians, which means they eat pollen and nectar from plants and usually have hairlike structures called setae that are prominent and branched in structure, the better for collecting pollen.

Yellowjackets are also not hornets, which are wasps in the genus *Vespa* and tend to have larger heads than yellowjackets. "North America has many native yellowjackets but no native hornets," says Kratzer. Though this does get a little confusing because there's one species of yellowjacket that is commonly known as the bald-faced hornet *(Dolichovespula maculata),* but it's not actually a hornet, scientifically speaking—and to complicate everything further, it's black-and-white in color, not yellow.

If you want to get technical, yellowjackets are half-inch-long insects in the genera *Vespula* and *Dolichovespula.* Like hornets, yellowjackets are a kind of wasp.

A social wasp, in fact, which means they create complex societies that work together to raise their young. This probably seems like a silly thing to point out, since we commonly picture bees, wasps, and hornets in hives or nests, but that arrangement is actually far from the norm.

"The vast majority of wasps are solitary," says Kratzer. "There are about 114,000 species of wasps in the world. And that's a very, very low estimate, because we're finding new ones at such a fast pace nowadays."

Solitary wasps are usually not much of a threat to humans, since they have nothing to defend. Rather, they wander around searching for tiny creatures to attack and devour, or sometimes to paralyze and lay their eggs inside.

As for the social wasps, Kratzer says that there are just 215 species known from Canada all the way down to Panama. And of those, only 22 species are yellowjackets. Although the yellowjackets are few in species, they are legion in range, number, and the ability to tick off humans.

In fact, one study polled 750 members of the general public for their attitudes about bees and wasps and learned that "bees are universally loved whilst wasps are universally despised." And that attitude is not new. Some 2,000 years ago, Aristotle penned this scathing review: "Hornets and wasps ... are devoid of the extraordinary features which characterize bees; this we should expect, for they have nothing divine about them as the bees have."

Clearly, it's time to set the record straight on yellowjacket awesomeness.

A Never-Ending Stream of Meatballs

There is a reason it seems like yellowjackets are messing with you. And it has to do with their tiny waists.

Like all members of the order Hymenoptera, which includes things like wasps, bees, and ants, yellowjackets can't eat solid food as adults. Rather, workers, males, and queens survive on a diet of sugar water. In the wild, the sweet stuff comes from flowers and the excretions of aphids and wasp larvae, but melted ice pops, fresh watermelon, and open cans of soda might as well be yellowjacket all-you-can-drink buffets. And this is why yellowjackets are always buzzing us at all our favorite outdoor events.

Yellowjacket nests reach their peak number of workers in the late summer and fall, which also happens to be when lots of flowering plants have already closed up shop and gone to seed. So the reason why it seems like the wasps are all up in your business is really a matter of economics—increased demand coupled with diminished supply equals increased yellowjacket encounters.

But while the search for sugar water brings us into frequent conflict with yellowjackets, it also overshadows two other kinds of foraging done by the insects on a regular basis. The first is the search for building materials.

Like their cousins known as paper wasps (genus *Polistes*), yellowjackets go out into the world and gnaw on various kinds of wood, combine that pulp with their spit, then haul it home to build nests with it. "They're like 3D printers," says Kratzer, "every time adding another little layer."

Preferences vary with the species, but yellowjackets can construct their papier-mâché nests hanging from a tree branch, in tree cavities, and even in abandoned rodent burrows underground. Of course, where humans are present, many more opportunities become available, and the insects can be found colonizing places like attics, windowsills, and the undersides of decks. This is yet another yellowjacket lifestyle choice that pits our species against one another.

But by far the coolest, most ferocious, and wholly unappreciated behavior wielded by yellowjackets everywhere is their ability to fan out over a 1,000-foot radius and slay every squiggly thing that moves. Yellowjackets have large, compound eyes that are especially good at detecting movement, and they use those peepers to pounce on everything from houseflies, horseflies, and disease-carrying mosquitoes to spiders, earthworms, and centipedes.

Intriguingly, yellowjackets don't normally sting their prey, but rather overpower their victims and then tear them limb from limb with their mandibles, or mouthparts. "They sort of mush their prey together into a ball that's easier to fly back home with," says Kratzer. Caterpillars are an especially prized prey item, probably because they're easily cut into meatballs and fed to rapacious larvae back at the nest. "Adult wasps are strict vegetarians," says Kratzer, "and the larvae are strict carnivores." Which means all of this butchery is done in the name of the yellowjackets' children.

Stake out a yellowjacket nest and you can actually watch as soldier after soldier returns bearing the mangled-up corpses of their victims—torsos, heads, and spheres of indeterminate meat matter all sourced from the world around you. Yellowjackets have also been known to tuck into the bodies of dead animals nearby and even extract tissue from open wounds on living horses.

One study revealed that as much as 95 percent of the animals dismantled and disappeared into the black hole of a social wasp nest were leaf-eating caterpillars—you know, the critters that plague farmers and devour our food before it can get to us.

"Wasps and ants as a group are the most important predators in all terrestrial environments," says Kratzer. "Without them, those prey species like caterpillars are going to completely defoliate their host plants, which isn't good for anybody."

At the end of the day, you can love or hate yellowjackets, just as you might any other animal in this book. But don't make the mistake of thinking these fantastic little Valkyries need to justify their existence to anyone.

Like Mexican free-tailed bats, bald eagles, or freshwater mussels, yellowjackets are just one of the innumerable paths life wandered down through deep time to arrive at the present. Each is an epic poem of trade-offs, persistence, and happenstance. Each is a grizzled survivor—unique and complete and imperfect. And if we allow any one of them to go extinct, then we've lost a song that can never be sung again.

And yet, somehow, every single day, scientists keep revealing new secrets about all the living things around us, from the hairy or scary to the scaly and extraordinary. And while some may look at a lamprey's maw or skunk's scent gland and recoil with fear or revulsion, I'm here to remind you that we always have another choice.

Choose wonder.

ACKNOWLEDGMENTS

I'D LIKE TO THANK every bug, book, teacher, writer, editor, scientist, and story that's pushed me, like a salmon, out of the ocean and upstream, over waterfalls and past hungry mouths, to arrive at this most sacred of places—the acknowledgments page.

This project would not have been possible without the backing and encouragement of National Geographic Books, as well as the careful hands of Susan Hitchcock, Tyler Daswick, Sanaa Akkach, Meredith Wilcox, Jen Hess, Jenny Miyasaki, and Becca Saltzman. It would not exist without the hustle of Jeff Kleinman at Folio Literary Management and Rachel Ekstrom Courage at Courage Literary. I also owe a great deal of gratitude to the generous support of the Alfred P. Sloan Foundation, as well as Doron Weber, Shriya Bhindwale, and Swan Griffith. I wrote the proposal for this book while completing a project fellowship at the Knight Science Journalism Program at MIT, and I can't thank Deborah Blum, Ashley Smart, and Tom Zeller, Jr., enough for that brief but life-changing experience. I also want to thank David Shiffman and Amy Panikowski for serving as my excellent scientific advisors for this manuscript. And thank you to Joel Sartore and Brian Skerry, who have brought this text to life with their stunning and inspiring photography, turned to elegant illustrations by artist Lisa Monias.

I've been fortunate to study under so many brilliant writers and teachers. In reverse order of appearance in my life: Joel Lovell, Jeanne Marie Laskas, Lee Gutkind, Lucy Fischer, Kerry Neville, Benjamin Slote, Matt Gross, Colleen Minerd, Alan Gillner, and Gina Cutlip—it's meant more than you know. Also, thanks to all my colleagues over the years who have read drafts of weird animal content that no doubt made very little sense at the time—from writing workshops at Allegheny College and the University of Pittsburgh Creative Nonfiction Program to the Think, Write, Publish fellowship at Arizona State University's Consortium for Science, Policy, and Outcomes and the Knight Science Journalism Program at MIT—each of you helped me get here.

In a similar vein, I've gotten to work with so, so many excellent editors over the years, each of whom has helped me become the writer I am today. (Everything you've ever read under my byline was made infinitely better by editors.) There are more than I can name, and I don't want to leave anyone out, but special commendation must go to Laura Helmuth for commissioning my very first piece—it was about porcupines!—and also to Melissa Mahony, Scott Dodd, and Torie Bosch for taking me under their wings at the very beginning of my career, and finally to Christine Dell'Amore, for whom I have had the honor to write more than any other editor in this business. When you find a good editor, never let them go.

I would also like to thank Rachel Gross, Maggie Koerth, Joe Hanson, Brandon Keim, and Glenn Oeland—each talented writers and science communicators who offered me their time, advice, and encouragement when they didn't have to. I have not forgotten.

If I'm leaving a few dozen excellent editors unnamed—but deeply appreciated—then the list of unmentioned scientists I'm shortchanging is easily in the hundreds. I remain in utter awe at the generosity, kindness, patience, and enthusiasm so many of them have granted me over these past 10 years of reporting. They have taken time to answer my questions while on vacation, while moving, while ill, and while dealing with personal tragedies. They have made me laugh, cry, and buckle beneath the wonder of it all.

To the scientists and experts quoted in this book, I hope I've done you and your animals proud. This book could not and should not exist without you.

Thank you to Jane Goodall, Marty Stouffer, Steve Irwin, Jack Hanna, Bill Nye, and David Attenborough. Thank you to Edward Abbey, John Steinbeck, and Annie Dillard. Thank you to Elizabeth Kolbert, Mary Roach, Michael Pollan, Cheryl Strayed, David Quammen, Carl Zimmer, Ed Yong, Jad Abumrad, and Robert Krulwich.

Thank you to my family and friends, who, let's be honest, have sat through more conversations about beetle biodiversity and carnivoran defensive secretions than most people would tolerate. Thank you to the WhatsApps, who keep me sane. Thank you to all my friends on Science Twitter, now on Bluesky, who actually probably could do with *more* musings about beetle biodiversity and carnivoran defensive secretions.

Thank you to Luanne, Joe, Albie, and Steve for *everything* you've done to help me get over the finish line.

Thank you to my dad, who taught me how to build a fire, hunt, and appreciate a cool mountain stream. Thank you to my mom, who dissected snakes, filled aquariums with tadpoles, and made plaster-of-Paris molds of bear footprints.

Thank you to Mark, June, and Huck, for snuggles, tickles, forts, Victory Royales, a wall full of animal artwork, and ultimately, for reminding me what all this is for.

Finally, thank you to Jess, who has been woken up so I could read her passages about corn, who has photographed pooping sloths, who has been nipped at by penguins, who has been rattled at by snakes, who has endured raccoon skulls as home decor, who has solo-parented so I could safari, and who has proudly loved and accepted me for the weirdo I have always been.

As Tom Robbins once wrote: "My love for you has no strings attached. I love you for free."

Original Photo

Hand-Silhouetted Image

Series of Filters

HOW THE ART WAS MADE

Each piece of art in this book began with photography, either by Joel Sartore or Brian Skerry, both longtime National Geographic photographers. Sartore is the founder of the Photo Ark, a 25-year documentary project to make portraits of all animals in human care around the world, and Skerry specializes in photographing marine wildlife and underwater realms.

The transformation of the photos into art was a multistage process, blending technology with creativity. Several carefully crafted scripts were applied at different stages, each using specialized filters to enhance, stylize, and refine the images. These digital techniques allowed textures to emerge, colors to shift, and details to take on a painted or illustrative quality, creating a fusion of realism and artistic interpretation. Layer by layer, the original images evolved—sometimes subtly, sometimes dramatically—until they reached their final form, a balance between photography and digital artistry.

—Lisa Monias

ILLUSTRATIONS CREDITS

All the illustrations in this book were created by Lisa Monias, based on Photo Ark photographs by Joel Sartore, except for the great white shark (page 228) and the Humboldt squid (page 284), which are based on photographs by Brian Skerry.

INDEX OF ANIMALS

Here we list, in the order they appear in the book, the common name of each species as well as the place where the original photograph was taken.

1: **Grizzly bear,** Sedgwick County Zoo, Wichita, Kansas
4: **North American mountain lion,** Rolling Hills Zoo, Salina, Kansas
7: **Mexican spotted owl,** New Mexico Wildlife Center, Española, New Mexico
8: **Grizzly bear,** Sedgwick County Zoo, Wichita, Kansas
16: **Southwestern bobcat,** Southwest Wildlife Conservation Center, Scottsdale, Arizona
18: **American buffalo,** Oklahoma City Zoo, Oklahoma City, Oklahoma
24: **Southwestern bobcat,** Southwest Wildlife Conservation Center, Scottsdale, Arizona
30: **Coyote,** Great Plains Zoo, Sioux Falls, South Dakota
36: **Geoffroy's spider monkey,** Gladys Porter Zoo, Brownsville, Texas
42: **Grizzly bear,** Sedgwick County Zoo, Wichita, Kansas
48: **Mexican free-tailed bat,** Night Wings Inc., Lubbock, Texas
54: **Mexican long-nosed armadillo,** Tulsa Zoo, Tulsa, Oklahoma
60: **Mexican wolf,** Endangered Wolf Center, Eureka, Missouri
66: **Mountain goat,** Cheyenne Mountain Zoo, Colorado Springs, Colorado
72: **Mountain lion**, ZooTampa at Lowry Park, Tampa, Florida
78: **North American beaver,** Nebraska Wildlife Rehab, Omaha, Nebraska
84: **North American porcupine,** Great Plains Zoo, Sioux Falls, South Dakota
90: **Orca,** SeaWorld San Diego, San Diego, California
96: **Ringtail,** Fort Worth Zoo, Fort Worth, Texas
102: **Star-nosed mole,** Kenora, Ontario, Canada
108: **Striped skunk,** Dunbar, Nebraska
114: **Virginia opossum,** Nebraska Wildlife Rehab, Omaha, Nebraska
120: **West Indian manatee,** The Dallas World Aquarium, Dallas, Texas

126: **White-tailed deer,** Gladys Porter Zoo, Brownsville, Texas
132: **Wolverine,** Zoo New York, Watertown, New York
138: **Northern spotted owl,** California Raptor Center, UC Davis, Davis, California
140: **American crow,** George Miksch Sutton Avian Research Center, Bartlesville, Oklahoma
146: **Bald eagle,** Sia: The Comanche Nation Ethno-Ornithological Initiative, Cyril, Oklahoma
152: **California condor,** Phoenix Zoo, Phoenix, Arizona
158: **Mexican spotted owl,** New Mexico Wildlife Center, Española, New Mexico
164: **Ruby-throated hummingbird,** Neale Woods Nature Reserve, Omaha, Nebraska
170: **Northwestern salamander,** A Rocha BC Centre, Surrey, British Columbia, Canada
172: **American alligator,** Kansas City Zoo & Aquarium, Kansas City, Missouri
178: **Axolotl,** Detroit Zoo, Royal Oak, Michigan
184: **Eastern indigo snake,** Toledo Zoo & Aquarium, Toledo, Ohio
190: **Gila monster,** Gladys Porter Zoo, Brownsville, Texas
196: **Eastern hellbender,** St. Louis Zoo, St. Louis, Missouri
202: **Leatherback sea turtle,** Bioko, Equatorial Guinea
208: **Cerbat Mountains rattlesnake,** Chiricahua Desert Museum, Rodeo, New Mexico
214: **Sonoran Desert toad,** Sedgwick County Zoo, Wichita, Kansas
220: **Snake River sockeye salmon,** Eagle Fish Hatchery, Eagle, Idaho
222: **Alligator gar,** Tennessee Aquarium, Chattanooga, Tennessee
228: **Great white shark,** Neptune Islands, South Australia, Australia
234: **Pacific lamprey,** MK Nature Center, Boise, Idaho
240: **Snake River sockeye salmon,** Eagle Fish Hatchery, Eagle, Idaho
246: **Indiana single flash firefly nymph,** Butterfly Pavilion, Westminster, Colorado
248: **Blue crab,** Louisiana
254: **Florida carpenter ant,** Urban Entomology Lab at the University of Florida, Gainesville, Florida
260: **Blue crayfish,** Oklahoma Aquarium, Jenks, Oklahoma
266: **Firefly,** Spring Creek Prairie Audubon Center, Denton, Nebraska
272: **Higgins eye pearlymussel,** Genoa National Fish Hatchery, Genoa, Wisconsin
278: **Harvester,** near Crosslake, Minnesota
284: **Humboldt squid,** Gulf of California, Mexico
290: **Monarch butterfly,** collected near Bennet, Nebraska
296: **Moon jelly,** Toledo Zoo, Toledo, Ohio
302: **Purple sea urchin,** Omaha's Henry Doorly Zoo and Aquarium, Omaha, Nebraska
308: **Hawaiian green darner,** Mt. Ka'ala Natural Area Reserve, Oahu, Hawaii
314: **Payson blonde tarantula,** Butterfly Pavilion, Westminster, Colorado
320: **Yellowjacket,** Audubon Insectarium, New Orleans, Louisiana

ABOUT THE CONTRIBUTORS

Jessica Bittel

JASON BITTEL is a wild animal most closely related to chimpanzees and bonobos. He is also a National Geographic Explorer, seasoned science journalist, and former project fellow at the Knight Science Journalism Program at the Massachusetts Institute of Technology. He once spent a season trapping invasive wild boars for the National Park Service. Bittel has written about science and animals for adult readers at *National Geographic,* the *New York Times,* the *Washington Post, Popular Science,* and *New Scientist,* among many other publications. He's also the author of four science books for kids: *Mountain* (2024), *The Frozen Worlds* (2023), *Animals Lost and Found* (2022), and *How to Talk to a Tiger ... and Other Animals* (2021). When he isn't sniffing sloth turds or professing his love for cockroaches, Bittel is howling at the moon with his wife and three kiddos in the woods outside of Pittsburgh. There's a good chance he's nursing a bad case of poison ivy right now. He also loves to cook and makes a mean cicada scampi, but the next batch won't be ready until 2036. To keep up with Bittel's writings and adventures, go to *www.bittelmethis.com.*

Cole Sartore

JOEL SARTORE is a photographer, author, conservationist, National Geographic Explorer, regular contributor to *National Geographic* magazine, and founder of the National Geographic Photo Ark. Over the years, his photographs have been seen on the pages of *Audubon, Sports Illustrated, Life, Smithsonian,* and the *New York Times.* He has published 12 Photo Ark books, seven for adults and five for children. He lives in Lincoln, Nebraska, with his wife, Kathy. Their three children and three grandchildren live nearby.

THE NATIONAL GEOGRAPHIC PHOTO ARK is a multiyear effort led by National Geographic Explorer and photographer Joel Sartore that aims to document the more than 25,000 species living in the world's zoos, aquariums, and wildlife sanctuaries; inspire action through education; and help protect wildlife by supporting on-the-ground conservation efforts.

Lisa Monias

LISA MONIAS is an illustrator, surface designer, and book creator. She is the author of several books, including *You Are Absolutely Awesome—Keep Rocking It!,* a compilation of short and fun inspirational sayings. When not lost in a world of drawing on her iPad or dreaming up her next project, she enjoys exploring the great outdoors with her dogs.

Since 1888, the National Geographic Society has funded more than 15,000 research, conservation, education, technology, and storytelling projects around the world. National Geographic Partners distributes a portion of the funds it receives from your purchase to National Geographic Society to support their mission to illuminate and protect the wonder of our world.

National Geographic Partners, LLC
1145 17th Street NW
Washington, DC 20036-4688 USA

Get closer to National Geographic Explorers and photographers, and connect with our global community. Join us today at nationalgeographic.org/joinus

For rights or permissions inquiries, please contact National Geographic Books Subsidiary Rights: bookrights@natgeo.com

Text copyright © 2026 Jason Bittel. Foreword copyright © 2026 Joel Sartore. Compilation copyright © 2026 National Geographic Partners, LLC. All rights reserved. Reproduction of the whole or any part of the contents without written permission from the publisher is prohibited.

NATIONAL GEOGRAPHIC and Yellow Border Design are trademarks of the National Geographic Society, used under license.

ISBN: 978-1-4262-2335-8

The authorized representative in the EU for product safety and compliance is Disney Trading B.V., Asterweg 15S, 1031 HL, Amsterdam, The Netherlands
email: DCP.DL-EU.bookscontact@disney.com

Printed in China

26/RRDH/1

EXPLORE THE WONDERS OF **THE PHOTO ARK**

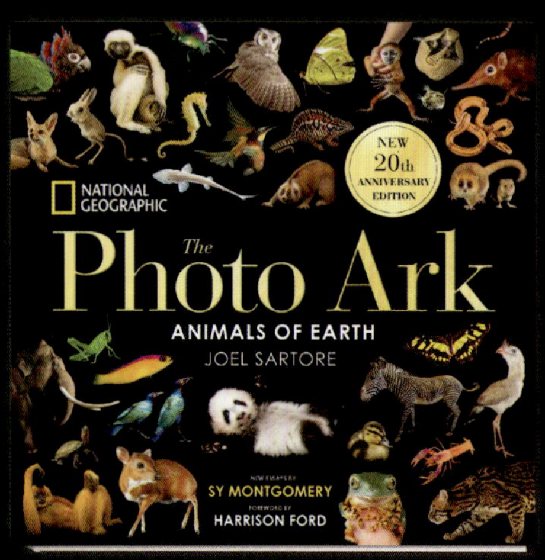

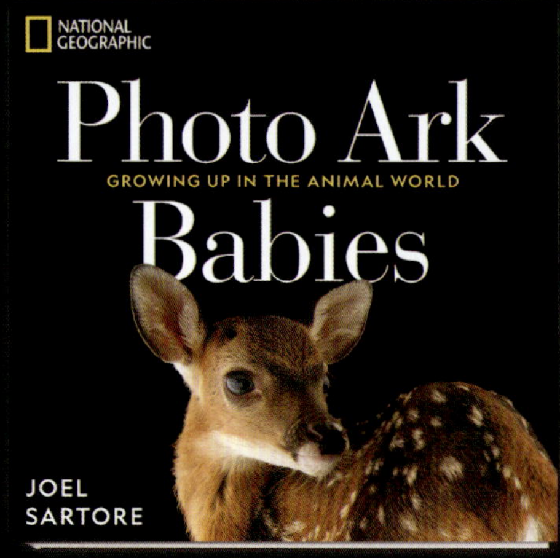

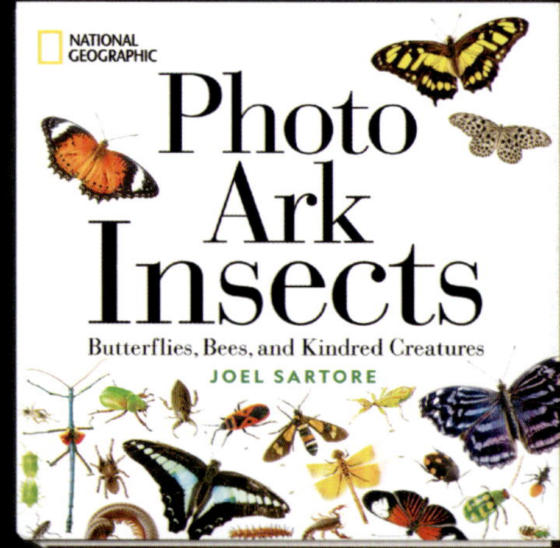